Sustainable Carbon Nanomaterials and their Applications

Online at: https://doi.org/10.1088/978-0-7503-6325-9

Sustainable Carbon Nanomaterials and their Applications

Edited by
Rafik Naccache
Department of Chemistry and Biochemistry and Centre for NanoScience Research (CeNSR), Concordia University, 7141 Sherbrooke St. West, Montreal, Quebec, Canada, H4B 1R6

Adedapo O. Adeola
Department of Chemistry and Biochemistry and Centre for NanoScience Research (CeNSR), Concordia University, 7141 Sherbrooke St. West, Montreal, Quebec, Canada, H4B 1R6

IOP Publishing, Bristol, UK

ISBN   978-0-7503-6325-9 (ebook)
ISBN   978-0-7503-6323-5 (print)
ISBN   978-0-7503-6326-6 (myPrint)
ISBN   978-0-7503-6324-2 (mobi)

DOI   10.1088/978-0-7503-6325-9

Version: 20251001

IOP ebooks

British Library Cataloguing-in-Publication Data: A catalogue record for this book is available from the British Library.

Published by IOP Publishing, wholly owned by The Institute of Physics, London

IOP Publishing, No.2 The Distillery, Glassfields, Avon Street, Bristol, BS2 0GR, UK

US Office: IOP Publishing, Inc., 190 North Independence Mall West, Suite 601, Philadelphia, PA 19106, USA

To Professor John A. Capobianco, whose lifelong curiosity about lanthanides, nano-materials, hybrid systems, and light has illuminated both science and the countless students and colleagues you have guided.

For more than forty years at Concordia University, you built a world-class lanthanide research program, co-founded the Concordia University Centre for NanoScience Research, and achieved the first observations of photon upconversion in $NaYF_4$ nanocrystals, discoveries that continue to shape bioimaging, phototherapy, and energy-conversion technologies. Your election as a Fellow of the Royal Society of Chemistry recognized not only your deep insight into rare-earth spectroscopy but also your conviction that rigorous science must serve society. Cited over 20000 times, your publications remain foundational references for anyone designing luminescent nanomaterials.

From harvesting sunlight in upconverting solar concentrators to creating photostable nanocatalysts that purify water, your work consistently points toward cleaner, more sustainable materials and processes. Yet numbers and breakthroughs tell only part of the story. What endures most is your generosity, teaching students to ask questions, urging collaborators to seek the unexpected glow, and proving by example that rigor and kindness can share the same laboratory bench.

For lighting the path and showing us how to walk it with integrity, this book is gratefully dedicated to you.

Professor Rafik Naccache
(Former graduate student of Professor Capobianco)
Editor

Contents

5   Nanoremediation of chemical pollutants using green carbon nanomaterials   5-1

Stephen Sunday Emmanuel and Ademidun Adeola Adesibikan

Preface

The 21st century continues to witness an unprecedented demand for materials that are not only high-performing but also environmentally sustainable. Amid growing concerns over climate change, pollution, and the depletion of natural resources, the field of nanotechnology has emerged as a promising frontier for addressing global sustainability challenges. Within this field, carbon nanomaterials have garnered significant attention due to their exceptional physical, chemical, and electronic properties and their versatile applications in energy, environmental remediation, and biomedical sciences.

This book, *Sustainable Carbon Nanomaterials and Their Applications*, is born out of the need to present a comprehensive and up-to-date exploration of how carbon nanomaterials can be designed, synthesized, and applied sustainably. In chapter 1, authored by the editors, Dr Adedapo Adeola and Professor Rafik Naccache, we present the motivation behind this book, linking the unique attributes of nano-materials with the principles of green chemistry and their growing relevance in the sustainability drive of the 21st century. This chapter sets the foundation by aligning nanoscience with sustainability goals, emphasizing the importance of life cycle thinking, resource efficiency, and environmentally benign processes in nanomate-rials research and application. Chapter 2, contributed by Professor Akeem Adeyemi Oladipo's team, focuses on the *classification and chemistry of carbon nanomaterials*, laying the scientific groundwork for the rest of the book. It covers their diverse forms, such as fullerenes, carbon nanotubes, graphene, and carbon dots, along with their structural features, electronic and optical properties, surface chemistry, and characterization techniques. The chapter concludes with a discussion of the multi-faceted applications of these materials and future research directions. In chapter 3, Professor Alberto Vomiero's team presents the *scalable and green synthesis of carbon dots*, along with their integration into sustainable technologies. This chapter discusses synthesis strategies, precursor effects, optical property tuning, and advanced characterization, culminating in an overview of practical applications in energy, sensing, and beyond. Chapter 4 emphasizes the importance of *computational and design-of-experiment (DOE) tools in optimizing the synthesis of sustainable carbon nanomaterials*. This chapter explores how advanced modeling and data-driven approaches are transforming nanomaterials research by reducing experimen-tal trial-and-error, conserving resources, and enabling precise material design and sustainability.

The subsequent chapters delve into various application domains of sustainable carbon nanomaterials. Environmental applications take center stage in chapter 5, where Dr Stephen Emmanuel's team discusses the use of *green carbon nanomaterials (GCNMs) for the remediation of organic and inorganic pollutants*. Emphasis is placed on the eco-efficiency, reusability, and real-world applicability of GCNMs, alongside a discussion of emerging research trends and gaps. In chapter 6, Professor Fernanda Maria Policarpo Tonelli's team explores the *biomedical applications* of carbon nanomaterials, including diagnostics, drug delivery, and therapeutic interventions. The chapter highlights how carbon nanomaterials can be engineered for enhanced biocompatibility, targeted delivery, and reduced toxicity, showcasing their growing potential in precision medicine. Chapter 7, by Dr Nsibande and Professor Patricia Forbes, addresses the *detection of environmental chemical contaminants* using eco-friendly carbon nanomaterials. This chapter underscores the importance of sensitive, selective, and sustainable sensors and sampling tools, which are crucial for monitoring pollutants and safeguarding environmental health.

Finally, the book concludes with chapter 8, authored by Dr Kayode Adegoke's team, which focuses on the *role of carbon-based materials in energy technologies*, covering catalytic processes for CO_2 conversion, biofuel production, green energy generation, and energy storage. The discussion highlights the central role carbon nanomaterials can play in transitioning toward low-carbon and renewable energy systems.

Overall, this book seeks to serve as a comprehensive resource for researchers, students, and professionals working in nanotechnology, environmental science, biomedical engineering, and materials chemistry. By bringing together foundational concepts, innovative applications, and sustainability-oriented methodologies, *Sustainable Carbon Nanomaterials and Their Applications* aims to inspire the development and deployment of carbon nanomaterials that contribute meaningfully to a cleaner, safer, and more sustainable future.

Professor Rafik Naccache and Dr Adedapo Adeola
Editors

Acknowledgements

The realization of this book would not have been possible without the dedication and expertise of the contributing authors, whose resilience and scholarly commitment shaped the content from concept to completion. We are especially grateful to the distinguished reviewers who provided critical feedback and expert insights during the book's developmental stages. Our sincere appreciation goes to Professor Akeem A. Oladipo of Eastern Mediterranean University, Professor Patricia Forbes of the University of Pretoria, Professor Alberto Vomiero of Luleå University of Technology, Professor Daniel Chua of the National University of Singapore, Professor José García Solé of Universidad Autónoma de Madrid, and Professor Giovanni Fanchini of Western University. Their thoughtful reviews were invaluable in ensuring the academic rigor and relevance of this work.

We also extend our heartfelt thanks to the exceptional team at IOP Publishing for their continued support, encouragement, and for the opportunity to bring this book to life. Their professionalism throughout the editorial process was instrumental to the success of this project. It is our hope that readers will find this book a valuable resource and a source of inspiration. We look forward to future opportunities to contribute to the advancement of nanoscience and its wide-ranging applications.

Professor Rafik Naccache and Dr Adedapo Adeola
Editors

Editor biographies

Rafik Naccache

Rafik Naccache obtained his PhD (2012) in chemistry at Concordia University in Quebec, Canada, working on lanthanide-doped upconverting nanoparticles for imaging applications. There, he was the recipient of the Distinguished Doctoral Dissertation Prize and the Governor General's Gold Medal for Technology, Industry, and the Environment. He subsequently carried out his Natural Sciences and Engineering Research Council of Canada (NSERC) postdoctoral training in nanobiophotonics at l'Institut National de la Recherche Scientifique, Quebec, Canada, developing terahertz sensing applications in nanobiophotonics. In December 2015, he accepted a tenure-track faculty position as a strategic hire in the Department of Chemistry and Biochemistry at Concordia University. Shortly thereafter, he was named a University Research Fellow and a Petro-Canada Young Innovator. He is currently an associate professor, the Director of the Centre for NanoScience Research, and a Concordia University Research Chair. His group's research focuses on the study of the fundamental properties of fluorescent carbon nanomaterials and hybrid nano-systems for the development of sensing, imaging, and catalysis applications. He has published over 70 manuscripts and has an h-index of 42 with over 9000 citations and 16 invited contributions, including reviews and book chapters, 80 conference presentations (48 invited, three keynotes, one plenary), and seven patents and disclosures of invention. He also possesses ten years of experience in pharmaceutical R&D obtained at Merck & Co., having been involved in materials characterization and drug development from 1998 to 2008. He still frequently consults for pharmaceutical companies, given his extensive expertise in the field.

Adedapo Adeola

Adedapo Adeola was awarded a PhD (2022) in chemistry at the University of Pretoria, South Africa, with a thesis focused on the development of advanced carbon-based materials for water treatment applications. He is passionate about green and sustainable approaches to environmental health and remediation. He is currently a postdoctoral fellow at the Centre for Nanoscience Research, focused on developing carbon nanomaterials and composites for contaminant detection and remediation. Adeola has published over 60 research articles in reputable journals, garnering more than 2000 citations, and has an h-index of 29 and an i10-index of 47. He has won several academic and research awards, including prestigious fellowships. Dr Adeola has contributed to several academic and research committees. He has held advisory, reviewer, and faculty roles at the Water Research Commission, South Africa;

Adekunle Ajasin University, Nigeria; and NSERC. His professional affiliations include membership in the Royal Society of Chemistry (RSC), the International Water Association (IWA), and the Society of Environmental Toxicology and Chemistry (SETAC). He has also held roles as a reviewer and editorial board member for major scientific journals and has gained industry experience as an environmental consultant in Southeast Asia and as a production engineer in the consumer goods sector in Nigeria. Dr Adeola's work reflects a strong commitment to sustainable innovation and interdisciplinary collaboration in advancing environmental health, material science and science education.

List of contributors

Kayode Adesina Adegoke
LAUTECH SDG 11 Sustainable Cities and Communities Research Group, Department of Pure and Applied Chemistry, Ladoke Akintola University, P. M. B. 4000, Ogbomoso, Nigeria

Dorcas Adenuga
Water Utilization Division, Department of Chemical Engineering, University of Pretoria, Pretoria, Private Bag X20, Hatfield, South Africa; Symbient Environmental Technologies, Mississauga, Ontario, Canada

Ademidun Adeola Adesibikan
Department of Chemistry, Faculty of Natural and Agricultural Sciences, University of Pretoria, Lynnwood Road, Hatfield, Pretoria 0002, South Africa

Abimbola O Aleshinloye
Department of Chemistry, Biochemistry and Physics, Florida Southern College, Lakeland Florida, 33801, USA

Hoda Ansari
Polymeric Materials Research Laboratory, Chemistry Department, Faculty of Arts and Science, TR North Cyprus, Eastern Mediterranean University, Famagusta via Mersin 10, Türkiye

Jahanara Sarker Ayesha
Department of Mechanical Engineering, Dhaka University of Engineering and Technology (DUET), Bangladesh

Geicielly da Costa Pinto
Federal University of São João del Rei, Centro-Oeste Campus, Divinópolis-MG, Brazil

Vitória de Oliveira Lourenço
Federal University of São João del Rei, Centro-Oeste Campus, Divinópolis-MG, Brazil

Adedipe Demilade
State Key Laboratory of Marine Pollution and Department of Chemistry, City University of Hong Kong, Hong Kong SAR, China

Stephen Sunday Emmanuel
Department of Industrial Chemistry, Faculty of Physical Sciences, University of Ilorin, P. M. B. 1515, Ilorin, Nigeria; Department of Chemical Sciences, Doornfontein Campus, University of Johannesburg, Doornfontein, Johannesburg, 2028, South Africa

Patricia B C Forbes
Department of Chemistry, Faculty of Natural and Agricultural Sciences, University of Pretoria, Lynnwood Road, Hatfield, Pretoria 0002, South Africa

Mustafa Gazi
Polymeric Materials Research Laboratory, Chemistry Department, Faculty of Arts and Science, TR North Cyprus, Eastern Mediterranean University, Famagusta via Mersin 10, Türkiye

Mahnoor Hassan
Department of Molecular Sciences and Nanosystems, Ca' Foscari University of Venice, Italy

Aliyu Ahmed Ibrahim
Polymeric Materials Research Laboratory, Chemistry Department, Faculty of Arts and Science, TR North Cyprus, Eastern Mediterranean University, Famagusta via Mersin 10, Türkiye

Kassa Belay Ibrahim
Department of Molecular Sciences and Nanosystems, Ca' Foscari University of Venice, Italy

Junior Kabanza Kalala
Polymeric Materials Research Laboratory, Chemistry Department, Faculty of Arts and Science, TR North Cyprus, Eastern Mediterranean University, Famagusta via Mersin 10, Türkiye

Md. Rifat Khandaker
Department of Chemical Engineering, Dhaka University of Engineering and Technology (DUET), Bangladesh

Pannan I Kyesmen
Department of Physics, Joseph Sarwuan Tarka University, Makurdi (formerly Federal University of Agriculture, Makurdi) P.M.B. 2373, Makurdi, Benue State, Nigeria

Elisa Moretti
Department of Molecular Sciences and Nanosystems, Ca' Foscari University of Venice, Italy

Sifiso A Nsibande
Department of Chemistry, Faculty of Natural and Agricultural Sciences, University of Pretoria, Lynnwood Road, Hatfield, Pretoria 0002, South Africa

Favour Ezinne Ogulewe
Polymeric Materials Research Laboratory, Chemistry Department, Faculty of Arts and Science, TR North Cyprus, Eastern Mediterranean University, Famagusta via Mersin 10, Türkiye

Akeem Adeyemi Oladipo
Polymeric Materials Research Laboratory, Chemistry Department, Faculty of Arts and Science, TR North Cyprus, Eastern Mediterranean University, Famagusta via Mersin 10, Türkiye

Flávia Cristina Policarpo Tonelli
Federal University of São João del Rei, Centro-Oeste Campus, Divinópolis-MG, Brazil

Fernanda Maria Policarpo Tonelli
Federal University of São João del Rei, Centro-Oeste Campus, Divinópolis-MG, Brazil

Tofik Ahmed Shifa
Department of Molecular Sciences and Nanosystems, Ca' Foscari University of Venice, Italy

Vinicius Marx Silva Delgado
Federal University of São João del Rei, Centro-Oeste Campus, Divinópolis-MG, Brazil

Christopher Santos Silva
Federal University of São João del Rei, Centro-Oeste Campus, Divinópolis-MG, Brazil

João Vitor Nunes
Federal University of São João del Rei, Centro-Oeste Campus, Divinópolis-MG, Brazil

Alberto Vomiero
Department of Molecular Sciences and Nanosystems, Ca' Foscari University of Venice, Italy; Department of Engineering Sciences and Mathematics, Luleå University of Technology, Sweden

Chapter 1

Introduction

Adedapo O. Adeola and Rafik Naccache

Carbon-based nanomaterials (CNMs), including graphene, carbon nanotubes, biochar, and carbon dots, are gaining prominence as sustainable materials for environmental, biomedical, and energy applications. Their exceptional surface properties, tunable chemistry, and biocompatibility make them ideal for applications such as pollutant removal, catalysis, energy storage, and sensor technologies. However, traditional CNM synthesis often involves toxic reagents and nonrenewable resources, conflicting with sustainability goals. This chapter explores how green chemistry principles, such as the use of renewable biomass feedstocks, solvent-free methods, and energy-efficient processes, are transforming CNM production. Emphasis is placed on recent advances in biomass-derived porous carbon materials, low-impact synthesis (e.g. hydrothermal carbonization, microwave pyrolysis), and life cycle thinking. This book bridges nano-science and green engineering by presenting CNMs as essential tools in achieving global sustainability targets and meeting 21st-century environmental and technological challenges.

1.1 Carbon-based nanomaterials and green chemistry principles

Carbon-based nanomaterials, such as graphene, carbon nanotubes, carbon dots, activated carbon, and fullerenes, have become increasingly vital in modern science due to their exceptional electrical, mechanical, thermal, and adsorptive properties [1, 2]. This class of nanomaterials/composites holds immense promise in diverse fields, including environmental remediation, energy storage, catalysis, and biomedical applications [3]. However, conventional methods for producing carbon nanomaterials often rely on nonrenewable precursors, hazardous chemicals, and energy-intensive procedures, raising significant environmental and techno-economic concerns [4]. To address these challenges, researchers are progressively adopting green chemistry principles to guide the sustainable development of carbon-based nanomaterials.

Green chemistry, a framework that emphasizes the minimization of toxic substances and waste throughout the life cycle of chemical products, offers practical strategies for making carbon nanomaterial synthesis safer and more sustainable [5]. Renewable biomass sources such as agricultural waste, cellulose, lignin, and other plant-derived materials are now being utilized as carbon precursors [6, 7]. These biogenic routes not only reduce dependency on fossil-based resources but also enable the transformation of waste into high-value functional nanomaterials [8]. The incorporation of green chemistry principles, such as the use of safer solvents, energy efficiency, waste prevention, and design for degradation, is essential when designing carbon-based nanomaterials for environmental, biomedical, and energy applications [9, 10]. Thermal or hydrothermal carbonization, microwave-assisted pyrolysis, and templating with natural salts or benign agents are among the green approaches gaining popularity for tailoring pore structures and surface functionalities in a controlled, eco-friendly manner [11, 12]. Additionally, surface modification of carbon nanomaterials using plant extracts or natural polymers further enhances their biocompatibility and application potential [13]. As environmental challenges related to water purification, pollution control, and energy transitions continue to persist [14], green approaches to the synthesis and application of carbon nano-materials will play a pivotal role in shaping a sustainable technological future.

1.2 Carbon-based nanomaterials and sustainability in the 21st century

In the current era of global challenges ranging from issues related to climate, energy, food security, health, and potable water supply, carbon-based nanomaterials have emerged as powerful tools for advancing sustainability. Nanomaterials exhibit exceptional physicochemical properties, high surface area, electrical conductivity, mechanical strength, and functional surface tunability, which are being harnessed to develop solutions in energy storage, pollution remediation, resource recovery, and green manufacturing [15]. Unlike many inorganic nanomaterials, carbon-based systems offer a unique advantage: they can often be derived from renewable or waste biomass sources, making them more compatible with circular and sustainable production approaches. One of the most transformative trends in carbon nano-material research is the shift toward low-cost and less toxic synthesis technologies using natural or waste-derived feedstocks [1]. Techniques such as hydrothermal carbonization, microwave-assisted pyrolysis, and chemical activation are enabling the scalable and eco-friendly production of biochar, graphene-like nanosheets, and carbon dots [4, 16]. These approaches are not only energy efficient but also eliminate the use of toxic solvents and harsh chemicals, aligning with key green chemistry principles and promoting zero-carbon footprints and sustainability [5]. The appli-cation of life cycle thinking in the synthesis and use of carbon nanomaterials helps evaluate their overall sustainability. By conducting life cycle assessments (LCAs) and environmental risk analyses, researchers can quantify emissions, resource consumption, and potential toxicity, ensuring that the nanomaterials developed are not only effective but also environmentally stable throughout their lifespan [17]. In essence, carbon-based nanomaterials represent a unique class of materials where the intersection of nanotechnology and green chemistry can yield high-performance engineered systems with reduced ecosystem footprints for a sustainable future.

1.3 The motivation for this book

The accelerating pace of population growth and industrialization, marked by a golden age of civilization, has simultaneously ushered in an urgent need for cleaner energy and sustainable resource use. The scientific community has been galvanized to a significant degree to explore innovative, scalable, and eco-friendly solutions to diverse global challenges. Among the various materials explored over the last two decades, carbon-based nanomaterials have distinguished themselves as some of the most promising candidates for addressing these challenges at the nexus of environmental remediation, biomedicine, and energy sustainability. With their tunable surface chemistry, high surface area, and remarkable structural versatility, carbon nanomaterials have been applied in a variety of fields [18].

The effectiveness of CNMs in capturing heavy metals, degrading pharmaceutical residues, and mitigating emerging contaminants in water systems has already been demonstrated in numerous studies. These materials are now being incorporated into next-generation membranes and catalytic platforms, offering highly selective, energy-efficient alternatives to conventional treatment technologies [19]. Simultaneously, the development of biomass-derived porous carbon materials has opened new avenues in sustainable energy storage. These carbon-based electrodes, often doped with nitrogen or other functional elements, are being engineered to support high-performance batteries and supercapacitors [20]. Their design not only meets technical demands for energy density and cycling stability but also aligns with sustainability principles by utilizing renewable feedstocks, minimizing raw material input, and promoting resource recovery. Adding further value to this class of nanomaterials is the growing field of bioinspired nanotechnology, where nature's strategies are mimicked to design smart, adaptive, and multifunctional carbon nanostructures [21]. These materials are now being fabricated through microwave-assisted, solvent-free, and low-energy routes, supporting the global transition toward greener manufacturing practices. Their applications span from nanosensors to controlled-release systems [22, 23], enabling more intelligent and sustainable technologies across sectors.

This book draws inspiration from the need to consolidate current knowledge and advance the conversation on the role of carbon-based nanomaterials in driving sustainable innovation. It is dedicated to recent advancements in the development and applications of sustainable carbon nanomaterials. The book comprises several chapters contributed by top international experts in the field on the topic of sustainable carbon nanomaterials and their applications. The goal is to ensure that this book takes a critical approach to the sourcing of sustainable materials, their synthesis and characterization, as well as applications spanning biomedicine and catalysis to green energy and environmental/analytical applications. The contributions will be comprehensive and will offer a seminal source for scientists already in the field, as well as those who intend to join this burgeoning research area. Carbon-based nanomaterials are no longer just materials of the future; they are critical tools for today's sustainability challenges. Their versatility, scalability, and ecological compatibility position them at the forefront of scientific and technological strategies aimed at securing a more resilient and resource-efficient 21st century.

References and further reading

[1] Rodoshi Khan N and Bin Rashid A 2024 Carbon-based nanomaterials: a paradigm shift in biofuel synthesis and processing for a sustainable energy future *Energy Convers. Manag.: X* **22** 100590

[2] Adeola A O and Forbes P B C 2024 Green carbon-based adsorbents for water treatment in Sub-Saharan Africa *Phys. Sci. Rev.* **9** 3563–77

[3] Asghar N, Hussain A, Nguyen D A, Ali S, Hussain I, Junejo A and Ali A 2024 Advancement in nanomaterials for environmental pollutants remediation: a systematic review on bibliometrics analysis, material types, synthesis pathways, and related mechanisms *J. Nanobiotechnol.* **22** 26

[4] Adeola A O, Duarte M P and Naccache R 2023 Microwave-assisted synthesis of carbon-based nanomaterials from biobased resources for water treatment applications: emerging trends and prospects *Front. Carbon* **2**

[5] Kurul F, Doruk B and Topkaya S N 2025 Principles of green chemistry: building a sustainable future *Discov. Chem.* **2** 68

[6] Adeola A O and Forbes P B C 2024 Two-dimensional carbon-based materials for sorption of selected aromatic compounds in water *Carbon Nanomaterials and their Composites as Adsorbents* ed J Tharini and S Thomas (Cham: Springer International Publishing) 247–60

[7] Villora-Picó J J, González-Arias J, Baena-Moreno F M and Reina T R 2024 Renewable carbonaceous materials from biomass in catalytic processes: a review *Materials (Basel)* **17**

[8] Saxena S, Moharil M P, Jadhav P V, Ghodake B, Deshmukh R and Ingle A P 2025 Transforming waste into wealth: leveraging nanotechnology for recycling agricultural byproducts into value-added products *Plant Nano Biol.* **11** 100127

[9] Slootweg J C 2024 sustainable chemistry: green, circular, and safe-by-design *One Earth* **7** 754–8

[10] Forbes P 2021 Green sample preparation methods in the environmental monitoring of aquatic organic pollutants *Curr. Opin. Green Sustain. Chem.* **31** 100500

[11] Senthilkumar A K, Kumar M, Kader M A, Shkir M and Chang J-H 2025 Unveiling the CO2 adsorption capabilities of carbon nanostructures from biomass waste: an extensive review *Carbon Capture Sci. Technol.* **14** 100339

[12] Adeola A O, Cui M and Naccache R 2023 Rhodamine B sequestration using acid-precipitated and microwave-treated softwood lignin: comparative isotherm, kinetics and thermodynamic studies *Environ. Technol. Innov.* **32** 103419

[13] Maghimaa M, Sagadevan S, Boojhana E, Fatimah I, Lett J A, Moharana S, Garg S and Al-Anber M A 2024 Enhancing biocompatibility and functionality: carbon nanotube-polymer nanocomposites for improved biomedical applications *J. Drug Deliv. Sci. Technol.* **99** 105958

[14] Shemer H, Wald S and Semiat R 2023 Challenges and solutions for global water scarcity *Membranes (Basel)* **13** 612

[15] Mahalakshmi D, Nandhini. J, Karthikeyan E, Karthik K K, Sujaritha J, Vandhana V and Lokeshwar R 2025 Carbon nanomaterials for emerging contaminant remediation: addressing pharmaceutical pollution in the water cycle with precision *Water Cycle* **6** 449–72

[16] Mahmood F *et al* 2025 A review of biochar production and its employment in synthesizing carbon-based materials for supercapacitors *Ind. Crops Prod.* **227** 120830

[17] Nizam N U M, Hanafiah M M and Woon K S 2021 A content review of life cycle assessment of nanomaterials: current practices, challenges, and future prospects *Nanomaterials (Basel)* **11** 3324

[18] Saleh M, Gul A, Nasir A, Moses T O, Nural Y and Yabalak E 2025 Comprehensive review of carbon-based nanostructures: properties, synthesis, characterization, and cross-disciplinary applications *J. Ind. Eng. Chem.* **146** 176–212

[19] Adeola A O, Paramo L, Fuoco G and Naccache R 2024 Emerging hazardous chemicals and biological pollutants in Canadian aquatic systems and remediation approaches: a comprehensive status report *Sci. Total Environ.* **954** 176267

[20] Qiu B, Hu W, Zhang D, Wang Y and Chu H 2024 Biomass-derived carbon as a potential sustainable material for supercapacitor-based energy storage: design, construction and application *J. Anal. Appl. Pyrolysis* **181** 106652

[21] Mundekkad D and Mallya A R 2025 Biomimicry at the nanoscale—a review of nanomaterials inspired by nature *Nano Trends* **10** 100119

[22] Macairan J-R, de Medeiros T V, Gazzetto M, Yarur Villanueva F, Cannizzo A and Naccache R 2022 Elucidating the mechanism of dual-fluorescence in carbon dots *J. Colloid Interface Sci.* **606** 67–76

[23] Adeola A O, Clermont-Paquette A, Piekny A and Naccache R 2024 Advances in the design and use of carbon dots for analytical and biomedical applications *Nanotechnology* **35** 012001

Chapter 2

Classifications and chemistry of carbon nanomaterials

Akeem Adeyemi Oladipo, Junior Kabanza Kalala, Abimbola O Aleshinloye, Favour Ezinne Ogulewe, Hoda Ansari, Aliyu Ahmed Ibrahim and Mustafa Gazi

This chapter focuses on the classification and chemistry of carbon nanomaterials, which have gained significant attention due to their unique properties and diverse applications. The chapter begins by defining carbon nanomaterials and providing an overview of their importance. It then delves into their classification based on structure and dimensionality, including fullerenes, carbon nanotubes, and graphene. The chemistry of carbon nanomaterials is explored, covering surface chemistry, electronic properties, optical properties, and chemical reactivity. Characterization techniques used to analyze these materials are discussed, followed by a comprehensive overview of their applications in various fields such as electronics, energy storage, biomedicine, and composites. The chapter concludes by summarizing key points and highlighting future perspectives and challenges in the field of carbon nanomaterials.

2.1 Introduction

Carbon nanomaterials, a class of materials composed of carbon atoms arranged in structures with at least one dimension in the nanometer range (1–100 nanometers), possess unique properties that distinguish them from their bulk counterparts. These exceptional properties arise from their nanoscale size and structure, making them highly versatile and promising for a wide range of applications. Carbon nanomaterials exhibit a diverse array of structural configurations, ranging from zero-dimensional (0D) fullerenes to one-dimensional (1D) carbon nanotubes and two-dimensional (2D) graphene [1–7]. This structural diversity, coupled with the unique properties of these nanomaterials, has garnered significant interest from researchers and industries alike.

The nanoscale dimensions of carbon nanomaterials allow for the emergence of quantum mechanical effects not observed in materials that lack nanoscale features

doi:10.1088/978-0-7503-6325-9ch2

[8–11]. These effects arise from the confinement of electrons within the nano-structures, which can lead to changes in their energy levels, resulting in altered electrical and optical properties. This phenomenon, known as quantum confine-ment, can give rise to unique electronic states and bandgaps, providing opportunities for the development of novel electronic and optoelectronic devices. Quantum confinement can occur in one, two, or three dimensions, depending on the size and shape of the nanomaterial. In 1D nanostructures [12–14], such as carbon nanotubes, the electrons are confined along one axis, leading to the formation of discrete energy levels that resemble those of atoms and molecules. This phenomenon is known as the quantum size effect and can result in the emergence of new electronic properties, such as metal–insulator transitions and the opening of bandgaps.

In 2D nanostructures [15], such as graphene, the electrons are confined in two dimensions, leading to the formation of a linear dispersion relation between energy and momentum. This linear dispersion relation results in exceptionally high electron mobility, making graphene a promising material for high-speed electronic devices, flexible electronics, and energy storage applications. In three-dimensional (3D) nanostructures, such as nanoparticles, the electrons are confined in all three dimensions, leading to the formation of discrete energy levels that are similar to those of atoms and molecules. This phenomenon is known as the quantum dot effect and can result in the emergence of new optical properties, such as luminescence and light emission. The quantum confinement of electrons in carbon nanomaterials can also give rise to other interesting phenomena, such as the formation of excitons, which are bound electron–hole pairs. Excitons can play a crucial role in the optical properties of carbon nanomaterials and can be used to create new types of lasers and light-emitting diodes.

In graphene, a single layer of carbon atoms arranged in a hexagonal lattice, the 2D confinement of electrons results in a linear dispersion relation between energy and momentum. This linear dispersion relation leads to exceptionally high electron mobility, making graphene a promising material for high-speed electronic devices, flexible electronics, and energy storage applications. In carbon nanotubes, cylin-drical structures composed of rolled-up graphene sheets, the 1D confinement of electrons can result in the formation of energy bands that are either metallic or semiconducting, depending on the chirality of the nanotube. This tunability of the bandgap makes semiconducting carbon nanotubes suitable for applications such as field-effect transistors, light-emitting diodes, and sensors.

The high surface-to-volume ratio of carbon nanomaterials provides a large number of active sites for interactions with other molecules or materials, enhancing their reactivity and functionality. This property is particularly advantageous for applications such as catalysis, adsorption, and sensing [9, 16, 17]. For example, functionalized carbon nanotubes can be used as supports for catalysts, improving their activity and selectivity. They can also be used as adsorbents for the removal of pollutants from water and air. The surface of carbon nanomaterials can be modified through various chemical and physical processes to introduce functional groups such as hydroxyl, carboxyl, and amino groups. These functional groups can be used to attach other molecules or materials to the surface of the carbon nanomaterial, creating new materials with tailored properties. For example, functionalized carbon nanotubes can be employed as building blocks for self-assembling nanostructures, which can be used in a variety of applications such as drug delivery and biosensing.

The specific properties of carbon nanomaterials depend on their structural characteristics, such as the arrangement of carbon atoms, the number of layers, and the presence of defects [18]. For instance, graphene exhibits exceptional mechanical strength, electrical conductivity, and thermal conductivity. Carbon nanotubes, cylindrical structures composed of rolled-up graphene sheets, can be either metallic or semiconducting depending on their chirality. Fullerenes, spherical structures composed of carbon atoms arranged in pentagons and hexagons, have unique electronic and optical properties.

This chapter presents a comprehensive overview of the classification, chemistry, and applications of carbon nanomaterials. It delves into their diverse structural configurations, unique properties, and potential applications. The fundamental principles governing the behavior of carbon nanomaterials, such as quantum confinement and surface chemistry, are explored. The implications of these properties for various technological advancements are extensively discussed. The chapter also emphasizes the challenges and opportunities associated with developing and utilizing carbon nanomaterials, highlighting ongoing research efforts and future prospects in this exciting field.

2.2 Classification of carbon nanomaterials

The classification of carbon nanomaterials provides a framework for understanding their unique properties and potential applications [17, 19, 20]. Based on structural characteristics such as the arrangement of carbon atoms, the number of layers, and defects, carbon nanomaterials can be classified as shown in figure 2.1. Additionally, their size and dimensionality influence their properties. Graphene, a 2D material with a hexagonal lattice structure, exhibits exceptional mechanical strength, electrical conductivity, and thermal conductivity due to its unique structure. Conversely, carbon nanotubes, a 1D material, have high aspect ratios and are suitable building blocks for conductive wires, sensors, and other applications.

By combining structural and size/dimensionality classifications, a more comprehensive understanding of carbon nanomaterials' diversity and potential applications can be achieved. For instance, single-walled carbon nanotubes (SWCNTs), cylindrical structures, can be either metallic or semiconducting depending on their chirality, which is determined by the angle at which the graphene sheet is rolled up. Multiwalled carbon nanotubes (MWCNTs), also 1D, possess multiple concentric layers of graphene sheets, offering enhanced mechanical strength and electrical conductivity compared to SWCNTs.

Furthermore, the classification of carbon nanomaterials can help identify trends and relationships between structure and properties. By studying the effects of different structural features, such as the number of layers in a carbon nanotube or the specific arrangement of carbon atoms in a fullerene, it is possible to gain insights into how these features influence the material's properties. This information can be

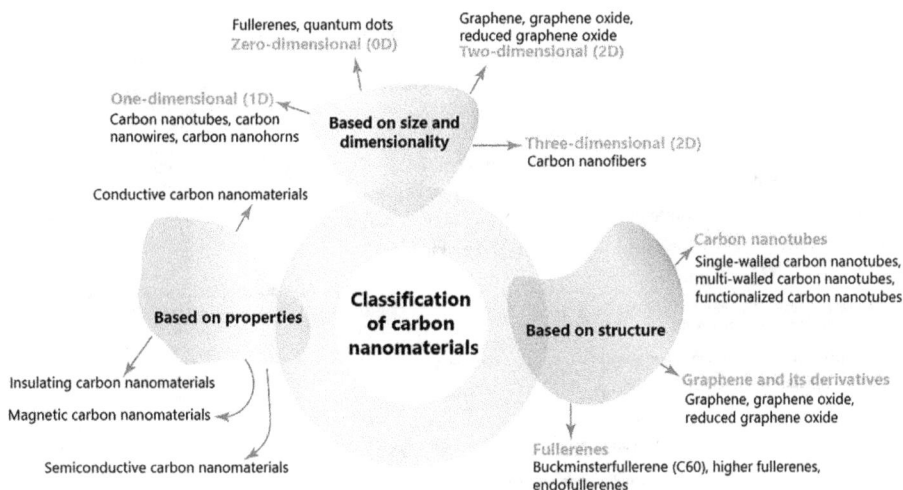

One-dimensional (1D)
Carbon nanotubes, carbon
nanowires, carbon nanohorns

Fullerenes, quantum dots
Zero-dimensional (0D)

Graphene, graphene oxide,
reduced graphene oxide
Two-dimensional (2D)

Based on size and
dimensionality

Three-dimensional (2D)
Carbon nanofibers

Conductive carbon nanomaterials

Carbon nanotubes
Single-walled carbon nanotubes,
multi-walled carbon nanotubes,
functionalized carbon nanotubes

Classification
of carbon
nanomaterials

Based on properties

Based on structure

Insulating carbon nanomaterials

Magnetic carbon nanomaterials

Graphene and its derivatives
Graphene, graphene oxide,
reduced graphene oxide

Semiconductive carbon nanomaterials

Fullerenes
Buckminsterfullerene (C60), higher fullerenes,
endofullerenes

Figure 2.1. Classification of carbon nanomaterials based on structural characteristics, dimensionality, and properties.

used to design and develop new carbon nanomaterials with tailored properties for specific applications.

2.2.1 Classification based on structure

2.2.1.1 Fullerenes: a unique class of carbon nanomaterials

Fullerenes, a distinctive class of carbon nanomaterials, are characterized by their spherical or cage-like structures [21–24]. These structures are formed by the arrangement of carbon atoms into pentagons and hexagons, creating a hollow, closed-shell configuration. The precise arrangement of these shapes is crucial for the stability and properties of fullerenes. The pentagons introduce curvature into the structure, allowing the carbon atoms to form a spherical or cage-like shape without significant strain. This curvature is essential for the stability of the fullerene structure, preventing the carbon atoms from forming linear chains or other less stable structures. The hexagons, on the other hand, provide a stable, planar arrangement of carbon atoms, contributing to the overall stability and rigidity of the fullerene. The arrangement of pentagons and hexagons in a fullerene can be described by its topology, a mathematical representation of the connectivity between the carbon atoms. The topology of a fullerene is determined by the number of pentagons and hexagons in the structure and how these shapes are connected.

The icosahedral topology is the most common fullerene topology, consisting of 12 pentagons and 20 hexagons arranged in symmetrical patterns (figure 2.2). However, other, more complex topologies exist. The specific arrangement of pentagons and hexagons in a fullerene determines its shape and size. The smallest and best-known fullerene is buckminsterfullerene (C_{60}), which is shaped like a soccer ball. Larger fullerenes with more pentagons and hexagons have more complex structures. The size and shape of a fullerene can influence its properties, such as its stability,

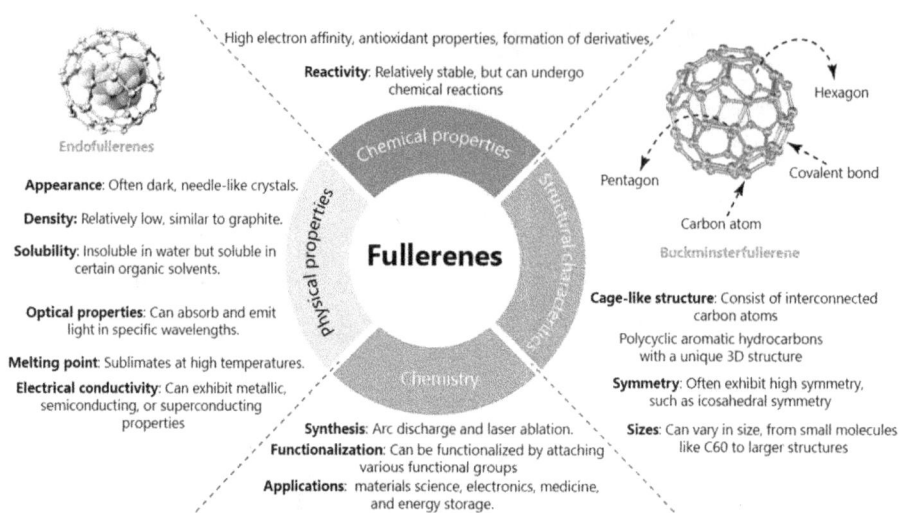

Figure 2.2. Chemical, physicochemical, and structural characteristics of fullerenes.

reactivity, and electronic behavior [21, 25, 26]. Fullerenes exhibit a range of unique properties. Their strong covalent bonds make them highly stable, resistant to degradation, and suitable for applications in harsh environments. Additionally, fullerenes possess unique electronic and optical properties not observed in other carbon materials. For example, they can exhibit nonlinear optical effects, which are useful for applications such as optical switching and data storage [26, 27].

Fullerenes can also be doped or functionalized to modify their properties and create new materials with tailored characteristics [28–30]. For instance, fullerenes' electrochemical properties can be significantly enhanced when they encapsulate various metals, such as lanthanides and alkaline earth metals. This functionalization leads to unique redox behaviors due to charge transfer between the metal and the fullerene cage [31]. Lanthanides, in particular, often donate three electrons to fullerenes, forming trivalent metallofullerenes. These functionalized fullerenes exhibit lower oxidation and reduction potentials compared to their empty counterparts [32, 33]. Interestingly, the redox potential is directly related to the ionic radius of the encapsulated metal: smaller metals tend to create fullerenes that are more easily oxidized and reduced than those with larger metals. Additionally, the size of the fullerene itself influences the number of possible redox reactions, with larger fullerenes exhibiting more options. While lanthanides and certain alkaline earth metals (calcium, magnesium) can donate two electrons to fullerenes, resulting in a wider electrochemical gap, some lanthanides, like cerium, lanthanum, and erbium, can accept up to six electrons, leading to relatively narrow bandgaps [34, 35]. Fullerenes can also be modified with organic groups, such as electron-rich moieties like amino groups, oligoethers, and crown ethers. This functionalization enhances the polarity of the fullerene derivatives, making them suitable for applications in perovskite solar cells [30].

Fullerenes can be synthesized through various methods [21, 22, 24], each with its advantages and disadvantages. One of the most common methods is arc discharge, which involves creating a high-temperature plasma between two graphite electrodes. The intense heat and energy of the plasma cause the graphite to vaporize, and the carbon atoms can then condense into fullerene structures. This method is relatively simple and can produce a wide range of fullerenes, including buckminsterfullerene and higher fullerenes. However, arc discharge can also produce a variety of impurities, such as amorphous carbon and other carbon nanomaterials, which may need to be removed through further purification steps. Another method for synthesizing fullerenes is laser ablation, which involves focusing a high-energy laser beam onto a graphite target. The laser vaporizes the graphite, and the carbon atoms can then condense into fullerene structures. Laser ablation can produce pure samples of fullerenes with a narrow size distribution, making it a desirable method for certain applications. However, laser ablation can be more expensive and less scalable than arc discharge.

Chemical vapor deposition is another method for synthesizing fullerenes. In chemical vapor deposition, a hydrocarbon gas is decomposed in a hot reactor, and the carbon atoms can then deposit onto a substrate to form fullerenes. Chemical vapor deposition can be used to produce fullerenes on a large scale and can also be used to synthesize other carbon nanomaterials, such as carbon nanotubes and graphene. However, chemical vapor deposition can be more challenging to control and can produce a wider range of products than arc discharge or laser ablation. The choice of synthesis method depends on the desired properties of the fullerenes. For example, arc discharge is often used to produce a wide range of fullerenes, including buckminsterfullerene and higher fullerenes. Laser ablation is particularly useful for producing pure samples of fullerenes with a narrow size distribution. Chemical vapor deposition can be used to produce fullerenes on a large scale and can also be used to synthesize other carbon nanomaterials, such as carbon nanotubes and graphene. The unique properties and versatility of fullerenes make them a promising area of research for scientists and engineers. Their potential applications span various fields, including electronics, materials science, and medicine. As research continues to advance, it is anticipated that fullerenes will be applied in even more innovative and exciting ways.

Key properties of fullerenes:
- **Spherical structure:** fullerenes have a unique spherical shape, which gives them distinct properties compared to other carbon nanomaterials.
- **Hollow interior:** the interior of fullerenes is hollow, which can be used to encapsulate other molecules or atoms.
- **Stability:** fullerenes are relatively stable due to their strong carbon-carbon bonds.
- **Unique electronic properties:** fullerenes exhibit unique electronic properties, including high electron affinity and low ionization energy.
- **Chemical reactivity:** the surface of fullerenes can be functionalized with various chemical groups, which can alter their properties and reactivity.

Types of fullerenes:

- **Buckminsterfullerene (C_{60})**: the most common fullerene, with a soccer ball-like structure. Its 60 carbon atoms are arranged in 12 pentagons and 20 hexagons, forming a highly symmetrical structure that resembles a truncated icosahedron [36–38]. This geometric shape is characterized by its 20 triangular faces, 30 edges, and 12 vertices. The arrangement of the carbon atoms in buckminsterfullerene is unique and creates a stable, spherical structure. This unique structure, combined with its exceptional properties, makes buckminsterfullerene a highly versatile and valuable nanomaterial. Its applications range from electronics and materials science to medicine and drug delivery [39–41].

- **Higher fullerenes**: fullerenes with more than 60 carbon atoms are known as higher fullerenes. These larger fullerenes have unique properties and can be used in various applications. For example, C_{70}, a fullerene with 70 carbon atoms, has a slightly elongated shape compared to C_{60}. This difference in shape can lead to different electronic and chemical properties. C_{80}, with 80 carbon atoms, has a more complex structure and can be used in applications such as organic electronics and drug delivery. Higher fullerenes can also be used to create fullerene-based polymers and nanostructures with unique properties.

- **Endofullerenes**: fullerenes that contain atoms or molecules trapped inside the hollow cage are known as endohedral fullerenes. The trapped atoms or molecules can be a variety of elements, including metals, rare gases, and even small clusters of atoms. The presence of these trapped atoms or molecules can significantly alter the properties of the fullerene, such as its electronic structure, magnetic properties, and reactivity [42, 43]. For example, endohedral fullerenes containing metal atoms can exhibit unique magnetic properties, making them potential candidates for applications in quantum computing and magnetic data storage. Endohedral fullerenes containing rare gases can be used as luminescent materials, with applications in lighting and display technologies. Additionally, endohedral fullerenes can be used to deliver drugs or other molecules to cells and tissues, as the trapped molecules can be released under specific conditions [44, 45]. Table 2.1 provides a comparative analysis of the unique features and structural characteristics of different fullerene types.

2.2.1.2 Carbon nanotubes: a promising class of nanomaterials

Carbon nanotubes are cylindrical structures composed of rolled-up graphene sheets, each of which is a single layer of carbon atoms arranged in a hexagonal lattice [46–48]. These unique carbon nanomaterials have garnered significant attention due to their exceptional properties (figure 2.3) and promising applications in various fields [46, 49, 50]. The cylindrical structure of carbon nanotubes can be either single-walled or multiwalled. SWCNTs consist of a single layer of graphene, while MWCNTs consist of multiple layers of graphene rolled up into concentric cylinders. The diameter of a carbon nanotube typically ranges from 1 to 100 nanometers, while the length can vary from a few nanometers to several micrometers.

Table 2.1. Comparison of the unique features of types of fullerenes.

Property/ feature	Buckminsterfullerene (C_{60})	Higher fullerenes	Endohedral fullerenes
Structure	Spherical, soccer ball-like	Spherical, with larger diameters	Spherical, with encapsulated atoms or molecules
Number of carbon atoms	60	Greater than 60	60 or more, with encapsulated atoms
Symmetry	High symmetry (truncated icosahedron)	Lower symmetry	May retain high symmetry if encapsulated atoms are centrally located
Chemical reactivity	Moderately reactive, can be functionalized	Generally less reactive than C_{60}	Reactivity can be influenced by the encapsulated atoms
Applications	Electronics, materials science, and drug delivery	Similar to C_{60}, but with the potential for unique properties	Electronics, materials science, drug delivery, catalysis
Key features	Highly stable, unique electronic properties	Larger size, the potential for new properties	Encapsulated atoms can alter properties

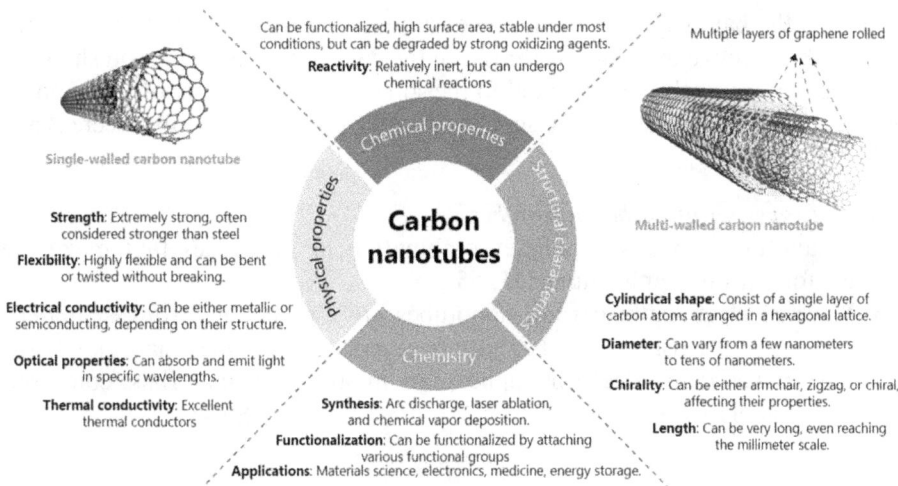

Figure 2.3. Chemical, physicochemical, and structural characteristics of carbon nanotubes.

The diameter of a carbon nanotube can be measured using various techniques, such as atomic force microscopy and transmission electron microscopy. The length of a carbon nanotube can be measured using techniques such as scanning electron microscopy and optical microscopy. The aspect ratio of a carbon nanotube is defined as its length divided by its diameter. Carbon nanotubes can have very high aspect ratios, which can make them useful for applications such as reinforcing composite materials and conducting electrical current [46, 51, 52].

The chirality of a carbon nanotube, determined by the angle at which the graphene sheet is rolled up, plays a crucial role in its electronic properties [53, 54]. Metallic carbon nanotubes exhibit no bandgap, while semiconducting carbon nanotubes possess a bandgap that can be tuned by controlling their chirality. The chirality of a carbon nanotube can be described by a pair of integers (n, m), where n and m are the numbers of unit vectors along two directions of the hexagonal lattice, as shown in figure 2.4. If $n = m$, the nanotube is an armchair, and it is metallic. If n-m is a multiple of three, the nanotube is zigzag, and it can be either metallic or semiconducting. If n and m are not equal or multiples of three, the nanotube is chiral, and it is always semiconducting.

The bandgap of a semiconducting carbon nanotube increases as its diameter decreases. This is because the confinement of electrons in a smaller-diameter nanotube becomes more pronounced, leading to a larger energy difference between the occupied and unoccupied states. As a result, the energy required to excite an electron from the valence band to the conduction band increases, resulting in a larger bandgap. The relationship between the bandgap and the diameter of a semiconducting carbon nanotube can be described by the following equation:

$$E_g = \frac{2h^2}{\pi^2 d^2 m_e},$$

where E_g is the bandgap, h is Planck's constant, d is the diameter of the nanotube, and m_e is the effective mass of an electron in the nanotube. This equation shows that the bandgap is inversely proportional to the square of the diameter of the nanotube. Therefore, as the diameter of a semiconducting carbon nanotube decreases, its bandgap increases.

2.2.1.2.1 Properties of carbon nanotubes

Carbon nanotubes possess a unique combination of properties that make them attractive for various applications [51, 55, 56]:

- **Mechanical strength:** carbon nanotubes are among the strongest materials known, with tensile strengths that are several times higher than that of steel. This makes them ideal for applications in structural materials, composites, and nanoelectronics.

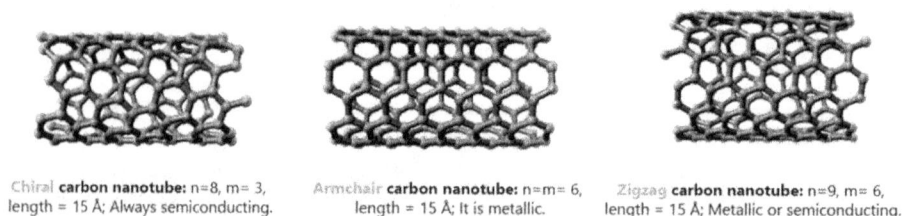

Chiral **carbon nanotube:** n=8, m= 3, length = 15 Å; Always semiconducting.   Armchair **carbon nanotube:** n=m= 6, length = 15 Å; It is metallic.   Zigzag **carbon nanotube:** n=9, m= 6, length = 15 Å; Metallic or semiconducting.

Figure 2.4. Chirality of carbon nanotubes.

- **Electrical conductivity:** metallic carbon nanotubes are excellent conductors of electricity, comparable to copper. This makes them suitable for use as interconnects in electronic devices and for applications in energy storage and generation.
- **Thermal conductivity:** carbon nanotubes have high thermal conductivity, making them promising materials for thermal management applications, such as heat sinks and cooling devices.
- **High surface area:** carbon nanotubes have a very high surface-to-volume ratio, making them excellent adsorbents and catalysts. This property is useful for applications in environmental remediation, energy storage, and chemical processes.
- **Optical properties:** carbon nanotubes can exhibit unique optical properties, such as photoluminescence and nonlinear optical effects. These properties make them promising materials for optoelectronic devices and sensors.

2.2.1.2.2 *Structure and types of carbon nanotubes*
- **SWCNTs** are cylindrical structures composed of a single layer of graphene rolled up into a tubular shape. The graphene sheet can be rolled up at different angles, which determines the chirality of the SWCNTs. The chirality of a SWCNT plays a crucial role in its electronic properties. Metallic SWCNTs are formed when the graphene sheet is rolled up at a particular angle, resulting in a structure that has no bandgap [57]. This means that there is no energy gap between the valence band and the conduction band in a metallic SWCNT, allowing electrons to move freely through the material. This makes metallic SWCNTs excellent conductors of electricity suitable for various electronic applications, such as interconnects and electrical wires [58].

  Semiconducting SWCNTs are formed when the graphene sheet is rolled up at a different angle, resulting in a structure that has a bandgap [59, 60]. The bandgap of semiconducting SWCNTs can be tuned by controlling their chirality. This tunability of the bandgap makes semiconducting SWCNTs promising materials for electronic devices such as field-effect transistors and light-emitting diodes. In a field-effect transistor [61, 62], the bandgap of the semiconducting material determines the on/off current ratio of the device. A larger bandgap leads to a higher on/off current ratio, which is desirable for digital electronics. In a light-emitting diode, the bandgap of the semiconducting material determines the wavelength of the emitted light. A smaller bandgap leads to light emission at longer wavelengths, while a larger bandgap leads to light emission at shorter wavelengths.
- **MWCNTs** are cylindrical structures composed of multiple layers of graphene rolled up into concentric cylinders [63]. These layers are arranged in a nested fashion, separating each layer from the adjacent layer by a small gap. The number of layers in a multiwalled carbon nanotube can vary widely, from a few to several hundred. The arrangement of these layers can be either concentric or helical. Concentric MWCNTs have layers arranged in a concentric pattern, similar to the layers of an onion. Helical MWCNTs

have layers arranged in a helical pattern, similar to the strands of a rope. The arrangement of the layers in a MWCNT can influence its properties, such as its mechanical strength, electrical conductivity, and thermal conductivity.

MWCNTs are generally more mechanically robust than SWCNTs [64, 65]. This is due to the presence of multiple layers, which provide additional strength and stability to the structure. MWCNTs are also more resistant to damage from external forces, such as mechanical stress or chemical attack. Another key advantage of MWCNTs is their high surface area. The multiple graphene layers in a multiwalled carbon nanotube create a large surface area, which can be beneficial for applications such as adsorption, catalysis, and energy storage. MWCNTs can be used as adsorbents to remove pollutants from water and air, as catalysts to enhance chemical reactions, and as electrodes in batteries and supercapacitors. The mechanical robustness and high surface area of MWCNTs make them attractive for various applications. For example, MWCNTs can be used to reinforce composite materials, such as polymers and metals, to improve their mechanical properties. They can also be used to fabricate sensors, actuators, and other electronic devices.

- **Functionalized carbon nanotubes** are carbon nanotubes that have been modified with various chemical groups or molecules [66, 67]. This modification process, known as functionalization, alters the properties of the nanotubes, making them suitable for a wider range of applications. The functional groups can be attached to the surface of the nanotubes through covalent or non-covalent bonds. Covalent functionalization involves the formation of new chemical bonds between the functional groups and the carbon atoms in the nanotube walls. This type of functionalization is typically more stable and can provide more durable modifications. However, covalent functionalization can be more challenging to achieve, as it often requires harsh reaction conditions or specialized reagents.

Non-covalent functionalization involves the formation of weak interactions, such as van der Waals forces or hydrogen bonding, between the functional groups and the nanotube surface. This type of functionalization is generally less stable than covalent functionalization, and the functional groups can be more easily removed under certain conditions. However, non-covalent functionalization can be easier to perform and can provide more reversible modifications. The choice of functional group and the method of functionalization can significantly affect the properties of the functionalized carbon nanotubes.

For example, the introduction of hydrophilic functional groups, such as hydroxyl or carboxyl groups, can improve the solubility and dispersibility of the nanotubes in water [66]. The introduction of hydrophobic functional groups, such as alkyl or aryl groups, can improve the compatibility of the nanotubes with organic solvents [68]. Functionalization can be achieved through various methods, including chemical reactions, sonication, plasma treatment, and click chemistry. The choice of functionalization method depends on the desired properties of the functionalized carbon nanotubes and the specific application.

2.2.1.3 Graphene: a marvel of modern materials

Graphene, a 2D sheet of carbon atoms arranged in a hexagonal lattice, has emerged as one of the most promising materials of the 21st century. Its unique properties have captivated scientists and engineers alike, leading to a surge of research and development in various fields [69–72]. One of the most remarkable features of graphene is its exceptional mechanical strength. It is the strongest material known to man, surpassing even steel and diamond. This incredible strength is due to the strong covalent bonds between the carbon atoms in the hexagonal lattice. Additionally, graphene is incredibly lightweight, making it an ideal material for applications where weight reduction is crucial.

Graphene is also an excellent conductor of both heat and electricity. Its high electron mobility allows for the rapid transfer of electrons, making it a promising material for electronic devices such as transistors and sensors. Furthermore, graphene's ability to conduct heat efficiently has applications in thermal management and energy storage. Another fascinating property of graphene is its transparency. At only one atom thick, graphene is almost completely transparent to visible light. This transparency, combined with its high electrical conductivity, makes it a potential candidate for flexible, transparent electronics. Graphene's unique properties (figure 2.5) have led to a wide range of potential applications. In electronics, it could revolutionize the field of flexible electronics, enabling the creation of devices that can be bent, folded, or even worn.

In energy storage, graphene-based materials could improve the performance of batteries and supercapacitors. In biomedical applications, graphene could be used for drug delivery, biosensors, and tissue engineering. Despite its immense potential, graphene still faces some challenges. One of the major hurdles is the difficulty in producing large-scale, high-quality graphene sheets. Additionally, the integration of graphene into existing manufacturing processes can be complex. However, researchers are actively working to address these challenges, and significant progress has been made in recent years.

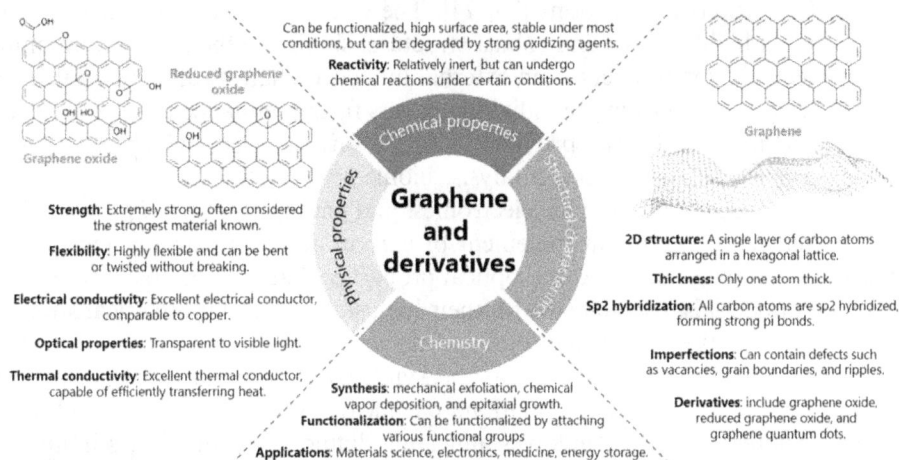

Figure 2.5. Chemical, physicochemical, and structural properties of graphene and its derivatives.

- **Graphene sheets** are single-layer sheets of carbon atoms arranged in a hexagonal lattice. They are the fundamental building blocks of other carbon nanomaterials, such as carbon nanotubes and graphene oxide. Graphene sheets exhibit exceptional properties, including high mechanical strength, electrical conductivity, and thermal conductivity. Due to these properties, graphene has attracted significant attention for potential applications in electronics, energy storage, and materials science.
- **Graphene oxide** is a derivative of graphene that is obtained by oxidizing graphene. The oxidation process introduces oxygen-containing functional groups, such as hydroxyl, epoxy, and carboxyl groups, onto the graphene surface. These functional groups alter the properties of graphene, making it more hydrophilic and soluble in water. Graphene oxide is often used as a precursor material for the production of reduced graphene oxide, which has properties similar to pristine graphene.
- **Reduced graphene oxide** is obtained by chemically or thermally reducing graphene oxide. The reduction process removes the oxygen-containing functional groups, restoring some of the properties of pristine graphene. However, reduced graphene oxide typically retains a higher degree of defects and impurities compared to pristine graphene. Despite these limitations, reduced graphene oxide is still a promising material for various applications due to its lower cost and ease of production compared to pristine graphene. Table 2.2 provides a general overview of the key features of different carbon nano-materials. Specific properties and applications can vary depending on the exact structure, size, and production method.

2.2.1.4 Carbon nano-onions

Carbon nano-onions are a unique type of carbon nanomaterial characterized by their onion-like structure, consisting of multiple concentric layers of carbon atoms arranged in a fullerene-like configuration. This distinctive arrangement imparts a wide range of exceptional properties that make carbon nano-onions promising materials for various applications [73, 74]. The concentric layers in carbon nano-onions create a highly stable and robust structure, similar to the layers of an onion, which protects the inner layers from external influences. The strong covalent bonds between the carbon atoms in the individual layers further enhance this stability. The concentric arrangement also provides a large surface area, which can benefit applications such as adsorption, catalysis, biomedical applications, environmental remediation, composite materials, electronics, and energy storage [75].

Additionally, the unique arrangement of carbon atoms in carbon nano-onions can give rise to novel electronic and optical properties that are not observed in other carbon nanomaterials. Combined with their high stability and mechanical strength, these properties make carbon nano-onions attractive materials for a wide range of applications. Carbon nano-onions can be synthesized through various methods, including arc discharge, laser ablation, chemical vapor deposition, and flame synthesis. Another approach involves annealing diamond nanoparticles in an inert atmosphere at high temperatures, which transforms them into small, spherical carbon nano-onion structures with six to eight layers. Ghalkhani and Sohouli [76]

Table 2.2. Comparison of key features of carbon nanomaterials.

Feature	Fullerenes	Carbon nanotubes	Graphene and derivatives	Carbon nanostructures
Structure	Spherical or cage-like structures	Cylindrical or tubular structures	Single-layer sheets of carbon atoms	Diverse shapes, including horns and fibers
Intrinsic properties	High symmetry, stability, and unique electronic properties	High aspect ratio, mechanical strength, and electrical conductivity	Excellent thermal and electrical conductivity, high surface area	Varying properties depending on the structure
Chemistry	Relatively inert, can be functionalized through chemical reactions	Can be functionalized through covalent or non-covalent modifications	Can be chemically modified through oxidation, reduction, and doping	Can be functionalized depending on the structure
Applications	Drug delivery, electronics, catalysis	Composites, electronics, energy storage	Electronics, sensors, energy storage	Composites, catalysis, energy storage
Challenges	Production cost, limited scalability	Production cost, alignment, dispersion	Scalable production, defect control, and toxicity	Production control, uniformity, characterization
Dimensionality	0D	1D	2D	Varying dimensions (can be 1D, 2D, or 3D)
Examples	Buckminsterfullerene (C_{60}), higher fullerenes	SWCNTs, MWCNTs	Graphene, graphene oxide, reduced graphene oxide	Carbon nanohorns, carbon nanofibers

successfully demonstrated this process by annealing diamond nanoparticles at 1650 °C for an hour and subsequently removing any amorphous carbon at 400 °C.

Another method for synthesizing carbon nano-onions is hydrothermal carbonization [77, 78]. Sang *et al* [78] used this technique to produce carbon nano-onions from citric acid in the presence of a KNO_3 catalyst. The yield and size of the resulting carbon nano-onions were influenced by factors such as the precursor concentration, reaction time, and solution pH. The formation of carbon nano-onions involves the multistep process illustrated in figure 2.6. Citric acid molecules initially dehydrate to form graphene quantum dots in the hydrothermal solution. These quantum dots then grow into graphene oxide layers, which stack to create graphite nanosheets. As the interface between the nanosheets and the solution increases, the sheets curve and roll into hollow polyhedra. Finally, through carbon rearrangement, these polyhedra transform into spherical carbon nano-onions, minimizing their surface energy.

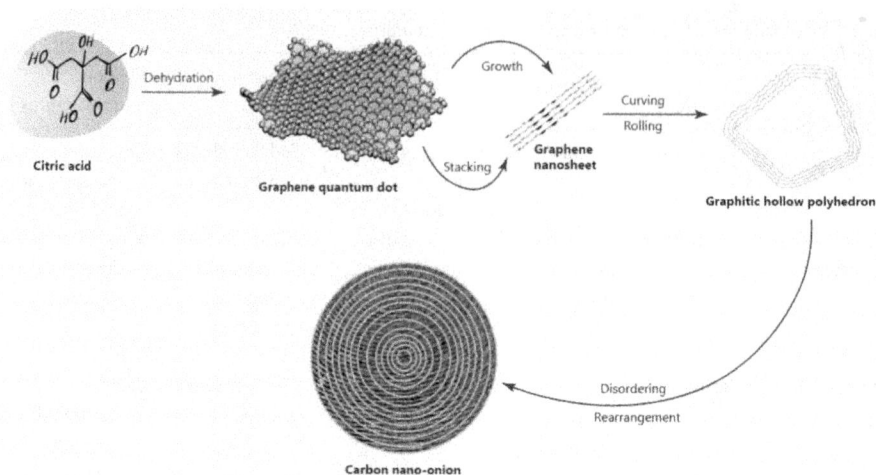

Figure 2.6. Schematic illustration of the hydrothermal synthesis of carbon nano-onion.

The chemistry of carbon nano-onions is primarily determined by the presence of defects and functional groups on their surface. Defects, such as vacancies, dislocations, or edges, can introduce reactive sites that can interact with other molecules or materials. Functional groups, such as hydroxyl, carboxyl, or amino groups, can be introduced onto the surface of carbon nano-onions through various chemical modifications, altering their properties and reactivity.

2.2.1.4.1 Key characteristics and properties of carbon nano-anions

- **Structure:** carbon nano-onions are composed of concentric fullerene-like shells, resembling an onion. The number of shells can vary, from a few to several hundred.
- **Size:** carbon nano-onions typically have a diameter of around 5–50 nm.
- **Stability:** their spherical structure and strong covalent bonding between the carbon atoms make carbon nano-onions highly stable, resistant to degradation, and compatible with various environments.
- **Mechanical strength:** carbon nano-onions exhibit exceptional mechanical strength and hardness, comparable to that of diamond. This makes them ideal for applications in structural materials and composites.
- **Thermal conductivity:** carbon nano-onions possess high thermal conductivity, making them suitable for thermal management applications, such as heat sinks and cooling devices.
- **Electrical conductivity:** carbon nano-onions can be either metallic or semi-conducting, depending on their structure. Metallic carbon nano-onions exhibit high electrical conductivity, while semiconducting carbon nano-onions have a bandgap that can be tuned by controlling their structure.
- **Optical properties:** carbon nano-onions can exhibit unique optical properties, such as photoluminescence and nonlinear optical effects. These properties make them promising materials for optoelectronic devices and sensors.

2.2.2 Classification based on size and dimensionality

Carbon nanomaterials, composed of carbon atoms arranged in structures with at least one dimension in the nanometer range, can be classified into three primary categories based on their size and dimensionality. Dimensionality, a measure of the number of dimensions in which a material extends, is a fundamental factor influencing the properties and applications of carbon nanomaterials [79]. The dimensionality of carbon nanomaterials significantly impacts their properties and potential applications. By comprehending the relationship between size and dimensionality, researchers can effectively design and engineer carbon nanomaterials with tailored properties for various fields, including electronics, materials science, and energy storage.

2.2.2.1 Zero-dimensional carbon nanomaterials

As their name implies, 0D carbon nanomaterials lack extended dimensions. Their nearly spherical structure, with a diameter typically in the nanometer range, results in a high surface-to-volume ratio. This configuration enables unique quantum mechanical effects due to the confinement of electrons within their small volume. Consequently, 0D carbon nanomaterials exhibit distinct properties compared to other carbon nanomaterials. The quantum confinement effects arising from this small volume can significantly alter their electronic and optical properties, including fluorescence, luminescence, and nonlinear optical effects. These nanomaterials often possess unique electronic properties, such as tunable bandgaps, high electron mobility, and strong light emission. Common examples of 0D nanomaterials include fullerenes and quantum dots [80].

2.2.2.2 One-dimensional carbon nanomaterials

One-dimensional nanomaterials, characterized by a significantly elongated structure with one dimension far exceeding the other two, possess unique properties that distinguish them from their 0D and 2D counterparts. Their cylindrical or tubular shape, often referred to as having a high aspect ratio, offers several advantages. The elongated structure facilitates efficient electron transport, while the large surface area it provides is beneficial for applications such as catalysis and adsorption. Moreover, the high aspect ratio enhances their mechanical properties, making them suitable for structural materials and composites.

This unique structure imparts distinctive properties, such as high electron mobility and tunable bandgaps. High electron mobility, the ability of electrons to move through a material with minimal resistance, is crucial for applications such as transistors and other electronic devices. Tunable bandgaps, which allow for the control of electron energy levels within the material, are essential for applications such as light-emitting diodes and solar cells. Prominent examples of 1D nanomaterials include carbon nanotubes and nanowires.

Physicochemical properties of 1D carbon nanomaterials:
- **High electron mobility:** 1D carbon nanomaterials, such as carbon nanotubes, have exceptionally high electron mobility, which makes them promising materials for electronic devices.

- **Tunable bandgaps:** the electronic properties of 1D carbon nanomaterials can be tuned by controlling their structure or composition, allowing for the creation of materials with specific bandgaps.
- **High mechanical strength:** 1D carbon nanomaterials, such as carbon nano-tubes, are known for their exceptional mechanical strength, which makes them ideal for applications in composites and structural materials.
- **High surface area:** the elongated structure of 1D nanomaterials gives them a high surface-to-volume ratio, which is beneficial for applications such as catalysis and adsorption.

2.2.2.3 Two-dimensional carbon nanomaterials

Two-dimensional carbon nanomaterials are a class of materials that are only a few atoms thick, yet possess extraordinary properties that make them highly promising for a wide range of applications. These materials are composed of atomically thin sheets of atoms arranged in a regular pattern, forming a crystalline lattice. The 2D nature of these materials allows for unique properties that are not observed in bulk materials. One of the most remarkable features of 2D carbon nanomaterials is their high surface-area-to-volume ratio. This means that they have a large surface area compared to their volume, which makes them ideal for applications such as catalysis, sensing, and energy storage. The high surface area of 2D carbon nano-materials provides numerous active sites for interactions with other materials, which can enhance their reactivity and functionality.

Another important feature of 2D carbon nanomaterials is their unique electronic properties [79, 81, 82]. These materials can exhibit high electron mobility, which means that electrons move through them very quickly. This makes them promising materials for electronic devices such as transistors and sensors. Additionally, 2D carbon nanomaterials can have tunable bandgaps, which means that their electrical conductivity can be controlled by changing their thickness or composition. This property is important for applications such as light-emitting diodes and solar cells. Graphene is the most well-known 2D carbon nanomaterial.

2.2.2.4 Three-dimensional carbon nanomaterials

This class of materials, with all dimensions within the nanometer range, is typically composed of interconnected networks of carbon nanostructures, such as graphene sheets or carbon nanotubes. This 3D structure imparts unique properties that distinguish it from its 2D or 1D counterparts. The interconnected nature of 3D carbon nanomaterials forms a complex network of pores and channels, customizable for specific applications through adjustments in synthesis conditions. This intricate structure enhances the mechanical properties of the material, providing a strong and rigid framework suitable for structural materials and composites. Moreover, the 3D structure improves thermal and electrical conductivity, making these materials promising for energy storage and electronics.

The 3D structure can also be manipulated to create hierarchical structures, enabling the development of materials with complex properties and functions. For instance, hierarchical structures can be designed with a high surface area and tunable pore size, which are essential for applications such as adsorption and filtration. In summary, the 3D structure of carbon nanomaterials offers unique properties that set these materials apart from their 2D or 1D counterparts. The interconnected network of nanostructures creates a versatile framework with customizable properties, making these materials promising for a wide range of applications. Common examples include graphene aerogels (lightweight, porous materials composed of interconnected graphene sheets) and carbon nanofibers (long, thin fibers of carbon atoms with a high aspect ratio that impart unique mechanical and electrical properties).

2.3 Chemistry of carbon nanomaterials

The chemistry of carbon nanomaterials is influenced by a complex interplay of factors, including the arrangement of carbon atoms, the presence of defects, and surface functionalization. The arrangement of carbon atoms in carbon nanomaterials can significantly affect their chemical properties [46, 83]. For example, graphene exhibits exceptional electrical conductivity due to the delocalized π-electrons that are free to move throughout the material. In contrast, carbon nanotubes, which are cylindrical structures composed of rolled-up graphene sheets, can be either metallic or semiconducting depending on their chirality, which is determined by the angle at which the graphene sheet is rolled up.

Defects (figure 2.7) in carbon nanomaterials, such as vacancies, dislocations, or grain boundaries, can profoundly impact their chemical properties [84]. Vacancies are missing atoms in the crystal lattice, while dislocations are defects in the

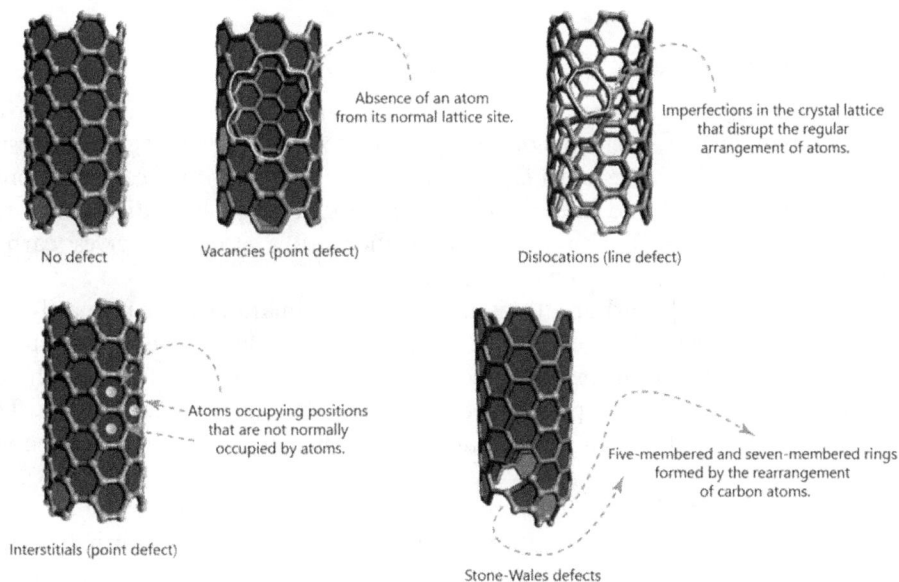

No defect   Vacancies (point defect)

Absence of an atom from its normal lattice site.

Dislocations (line defect)

Imperfections in the crystal lattice that disrupt the regular arrangement of atoms.

Interstitials (point defect)

Atoms occupying positions that are not normally occupied by atoms.

Stone-Wales defects

Five-membered and seven-membered rings formed by the rearrangement of carbon atoms.

Figure 2.7. Schematic representation of common defects in carbon nanomaterials.

arrangement of atoms. Grain boundaries are interfaces between different crystal orientations within a material. These defects can introduce new energy levels into the electronic structure of the material, which can alter its electrical conductivity, optical properties, and chemical reactivity. For example, vacancies can create localized states within the bandgap of a semiconductor, which can act as electron traps or recombination centers. Dislocations can introduce strain fields into the material, which can affect its mechanical properties and electronic structure. Grain boundaries can also introduce new electronic states and defects, which can influence the material's electrical conductivity and chemical reactivity [85, 86].

In addition to altering the electronic structure of carbon nanomaterials, defects can also create active sites on the surface of the material. These active sites can serve as catalysts for chemical reactions, promoting the formation of new products. For example, defects in graphene can create active sites for the reduction of oxygen, which is an important reaction in fuel cells and other energy storage devices [87, 88]. Defects can also be used to functionalize carbon nanomaterials with specific chemical groups, which can alter their properties and reactivity.

Surface functionalization is another important factor that influences the chemistry of carbon nanomaterials. By introducing functional groups such as hydroxyl, carboxyl, or amino groups onto the surface of carbon nanomaterials, it is possible to modify their properties and reactivity. For example, functionalized carbon nanomaterials can be used to attach other molecules or materials to the surface, creating new materials with tailored properties. The combination of these factors, including the arrangement of carbon atoms, the presence of defects, and the surface functionalization, gives carbon nanomaterials a wide range of chemical properties that can be tailored for specific applications.

2.3.1 Bonding and structure

The unique structural and electronic properties of carbon nanomaterials arise from the specific bonding arrangements of their carbon atoms. These arrangements result in the formation of strong, covalent bonds between neighboring carbon atoms, creating a highly stable and interconnected network. The sp^2 hybridization of the carbon atoms, combined with the delocalization of π electrons, gives carbon nanomaterials their distinctive characteristics.

- **sp^2 hybridization**: carbon atoms in carbon nanomaterials are primarily sp2 hybridized. This means that one 2s orbital and two 2p orbitals of the carbon atom combine to form three equivalent sp^2 hybrid orbitals, which are arranged in a trigonal planar geometry at $120°$ angles to each other. The remaining unhybridized 2p orbital is perpendicular to the plane of the sp^2 orbitals.
- **π-electron delocalization**: the unhybridized 2p orbitals of the carbon atoms in carbon nanomaterials overlap with neighboring p orbitals to form π bonds. These π bonds are delocalized over the entire structure, creating a cloud of electrons that is free to move throughout the material. This delocalization is

responsible for many of the unique properties of carbon nanomaterials, such as their electrical conductivity and thermal conductivity.

- **Aromatic nature**: the delocalized π electron system in carbon nanomaterials gives them an aromatic character. Aromatic molecules are characterized by their cyclic structure, planar geometry, and the presence of a delocalized π electron system that follows Hückel's rule ($4n + 2$ π electrons, where n is an integer). This aromatic character contributes to the stability and reactivity of carbon nanomaterials.

2.3.2 Surface chemistry

The surface of carbon nanomaterials, being a highly reactive region, is susceptible to various chemical and physical modifications. These modifications can involve the introduction of functional groups, such as hydroxyl, carboxyl, and amino groups, through processes such as oxidation, reduction, or grafting. The presence of these functional groups significantly alters the surface properties of carbon nanomaterials, influencing their interactions with other molecules and materials. For instance, hydroxyl groups can enhance the hydrophilicity of carbon nanomaterials, making them more compatible with aqueous environments. Carboxyl groups can impart acidic properties, enabling the formation of ionic bonds with other molecules. Amino groups, on the other hand, can introduce basic properties, facilitating interactions with positively charged species.

By carefully controlling the type and density of functional groups introduced onto the surface of carbon nanomaterials, it is possible to tailor their properties for specific applications. For example, functionalized carbon nanomaterials can be used as adsorbents to remove pollutants from water and air, as catalysts for chemical reactions, and as building blocks for self-assembling nanostructures. Surface functionalization of carbon nanomaterials can also be achieved through physical methods, such as plasma treatment or ion bombardment. These methods can introduce defects or modify the surface morphology of the carbon nanomaterial, which can also affect its chemical properties.

Plasma treatment involves exposing the carbon nanomaterial to plasma, which is a highly energetic gas composed of ions, electrons, and neutral atoms or molecules. The interaction between the plasma and the carbon nanomaterial can introduce defects, such as vacancies or dangling bonds, on the surface of the material. These defects can act as active sites for chemical reactions or can be used to functionalize the surface with specific chemical groups. Ion bombardment involves bombarding the carbon nanomaterial with a beam of ions, which can cause physical damage to the surface of the material. This damage can create defects or modify the surface morphology, which can also affect its chemical properties. The choice of physical method for surface functionalization depends on the desired properties of the carbon nanomaterial. Plasma treatment is often used to introduce defects or functional groups onto the surface of the material, while ion bombardment is more commonly used to modify the surface morphology. It is also possible to combine physical and chemical methods to achieve the desired surface functionalization.

2.3.3 Electronic properties

The electronic properties of carbon nanomaterials are determined by the arrangement of carbon atoms in their structure. Graphene, a single layer of carbon atoms arranged in a hexagonal lattice, exhibits exceptional electrical conductivity due to its delocalized π electrons, which are free to move throughout the material. Carbon nanotubes, cylindrical structures composed of rolled-up graphene sheets, can be either metallic or semiconducting depending on their chirality, which is the angle at which the graphene sheet is rolled up. Metallic carbon nanotubes have no bandgap, while semiconducting carbon nanotubes have a bandgap that can be tuned by controlling their chirality. This tunability of the bandgap makes semiconducting carbon nanomaterials promising materials for electronic devices [89, 90]. Doping carbon nanomaterials with appropriate impurities can further control their electrical conductivity and create materials with specific properties.

The electronic properties of carbon nanomaterials can also be influenced by defects, which can introduce new energy levels into the material's electronic structure. For example, vacancies can create localized states within the bandgap of a semiconductor, which can act as electron traps or recombination centers. Dislocations and grain boundaries can also introduce new electronic states and defects, influencing the material's electrical conductivity. Carbon nanomaterials exhibit a wide range of electrical conductivity properties, from highly conductive to insulating. This diversity in conductivity is a key factor that contributes to the versatility of carbon nanomaterials in various applications.

2.3.3.1 Conductivity of carbon nanomaterials

- **Fullerenes:** fullerenes are generally poor conductors of electricity due to their closed-shell structure, which limits the movement of electrons. However, certain fullerenes, such as buckminsterfullerene (C_{60}), can exhibit metallic or semiconducting behavior under specific conditions.
- **Carbon nanotubes:** carbon nanotubes can exhibit either metallic or semiconducting behavior, depending on their chirality. Metallic carbon nanotubes have no bandgap, allowing them to conduct electricity efficiently. Semiconducting carbon nanotubes have a bandgap, similar to semiconductors, which can be tuned by controlling their chirality. This tunability of the bandgap makes semiconducting carbon nanotubes promising materials for electronic devices such as field-effect transistors and light-emitting diodes.
- **Graphene:** graphene is an excellent conductor of electricity due to its unique 2D structure and the delocalized nature of its π electrons. It has a high electron mobility, which is a measure of how quickly electrons can move through a material. This makes graphene a promising material for high-performance electronic devices.

2.3.4 Optical properties

Carbon nanomaterials exhibit a wide range of optical properties, including strong light absorption, emission, and nonlinear optical effects. These properties are influenced by the size, shape, and structure of the carbon nanomaterial. For example, carbon nanotubes can exhibit strong light absorption in the ultraviolet and visible regions of the spectrum, while graphene can exhibit strong light absorption in the near-infrared region. The optical properties of carbon nanomaterials can be tailored by controlling their size, shape, and structure. For example, the diameter of a carbon nanotube can affect its optical properties, with smaller-diameter nanotubes exhibiting stronger light absorption.

The arrangement of carbon atoms in a graphene sheet can also influence its optical properties. For example, graphene nanoribbons with different edge configurations can exhibit different optical properties. In addition to their linear optical properties, carbon nanomaterials can also exhibit nonlinear optical effects. These effects occur when the optical properties of a material change with the intensity of the incident light. Carbon nanomaterials can exhibit a variety of nonlinear optical effects, such as second-harmonic generation, third-harmonic generation, and two-photon absorption. These effects can be useful for applications such as optical switching, optical data storage, and laser technology. The optical properties of carbon nanomaterials can be characterized using various techniques, such as UV–visible spectroscopy, Raman spectroscopy, and photoluminescence spectroscopy. These techniques can provide information about the absorption, emission, and nonlinear optical properties of carbon nanomaterials.

2.3.5 Chemical reactivity

Carbon nanomaterials exhibit a wide range of chemical reactivity, depending on their specific structure and surface properties. This reactivity can be influenced by factors such as the presence of defects, surface functionalization, and the interaction with other molecules or materials. One of the key factors that determines the chemical reactivity of carbon nanomaterials is the presence of defects in the carbon lattice. Defects, such as vacancies, dislocations, or edges, can create reactive sites that can interact with other molecules. These reactive sites can be more susceptible to chemical reactions than the perfect, defect-free regions of the carbon nanomaterial. For example, defects in graphene can act as catalysts for various chemical reactions, such as the decomposition of hydrogen peroxide or the reduction of carbon dioxide.

The type of defect can also influence the chemical reactivity of carbon nanomaterials. For example, vacancies, which are missing carbon atoms in the lattice, can create highly reactive sites. Dislocations, which are defects in the crystal structure, can also create reactive sites, but the reactivity of dislocations can depend on their orientation and the type of dislocation. Edges, which are the boundaries between different graphene layers in a carbon nanotube or graphene sheet, can also be reactive sites. The reactivity of edges can be influenced by their orientation and the presence of functional groups.

Another important factor that influences the chemical reactivity of carbon nanomaterials is their surface functionalization. The introduction of functional groups, such as hydroxyl, carboxyl, or amino groups, can significantly alter the reactivity of carbon nanomaterials. These functional groups can create new reactive sites or modify the electronic properties of the nanomaterial, making it more or less reactive toward certain molecules. For example, the introduction of hydroxyl groups can increase the reactivity of carbon nanomaterials toward oxidizing agents, while the introduction of carboxyl groups can increase their reactivity toward reducing agents.

The reactivity of carbon nanomaterials can also be influenced by interactions with other molecules or materials. For example, carbon nanomaterials can adsorb various molecules from the environment, which can affect their reactivity. The adsorbed molecules can create new reactive sites or block existing reactive sites, thereby altering the chemical reactivity of the carbon nanomaterial. Additionally, carbon nanomaterials can interact with other nanomaterials, such as metal nano-particles or other carbon nanomaterials, to form composite materials with new and unique properties. These composite materials can exhibit different chemical reac-tivities than the individual components. Carbon nanomaterials can interact with a variety of molecules and materials, including gases, liquids, and other nanomate-rials. These interactions can lead to a range of chemical reactions, such as oxidation, reduction, adsorption, and catalysis.

- **Oxidation:** carbon nanomaterials can be oxidized by exposure to oxidizing agents, such as oxygen, ozone, or hydrogen peroxide. Oxidation can introduce oxygen-containing functional groups, such as hydroxyl and car-boxyl groups, onto the surface of the nanomaterial.
- **Reduction:** carbon nanomaterials can be reduced by exposure to reducing agents, such as hydrogen or hydrazine. Reduction can remove oxygen-containing functional groups from the surface of the nanomaterial, creating new reactive sites.
- **Adsorption:** carbon nanomaterials can adsorb various molecules from the environment. This adsorption can result from physical forces, such as van der Waals interactions, or from chemical bonding between the carbon nano-material and the adsorbed molecule.
- **Catalysis:** carbon nanomaterials can act as catalysts for various chemical reactions. The catalytic activity of carbon nanomaterials can be influenced by their structure, surface properties, and the presence of defects or functional groups.

The chemical reactivity of carbon nanomaterials can be characterized using various techniques, such as x-ray photoelectron spectroscopy (XPS), Fourier transform infrared (FTIR) spectroscopy, and Raman spectroscopy. These techniques can provide information about the functional groups present on the surface of the nanomaterial and the changes in the chemical bonds that occur during chemical reactions. The chemical reactivity of carbon nanomaterials is an important factor that determines their potential applications. For example, the reactivity of carbon nanomaterials can be exploited for applications in catalysis, environmental reme-diation, and energy storage. By understanding the chemical reactivity of carbon nanomaterials, it is possible to design and develop new materials with tailored properties for specific applications.

2.4 Applications of carbon nanomaterials

Carbon nanomaterials, with their exceptional properties, have found diverse applications across various fields, as illustrated in figure 2.8 [75, 91–94]. In electronics, their exceptional electrical conductivity and mechanical strength make them ideal for use as conductive wires, transistors, and sensors. Their high surface area makes them ideal for energy storage applications, such as batteries and supercapacitors, where they enhance energy density and charge/discharge rates. Carbon nanomaterials can also be used as electrodes in fuel cells, where they improve the efficiency of the electrochemical reactions.

In biomedical engineering, carbon nanomaterials are employed for drug delivery, tissue engineering, and biosensing due to their biocompatibility and ability to interact with biological molecules. Their mechanical properties make them valuable in composite materials, improving strength, stiffness, and conductivity. Carbon nanomaterials can reinforce polymers, metals, and ceramics, creating materials with enhanced performance in various applications. In environmental remediation, carbon nanomaterials can be used to remove pollutants from water and air due to their adsorption capabilities.

Their high surface area and unique chemical properties allow them to efficiently adsorb pollutants such as heavy metals, organic compounds, and dyes. Additionally, carbon nanomaterials can be used as catalysts in various chemical processes, enhancing reaction rates and improving selectivity. Their catalytic properties can be tailored by modifying their surface chemistry or introducing functional groups. Table 2.3 provides a general overview of the chemistries, properties, structures, and synthesis methods of common carbon nanomaterials. The specific properties and applications can vary depending on the size, shape, and functionalization of the nanomaterial.

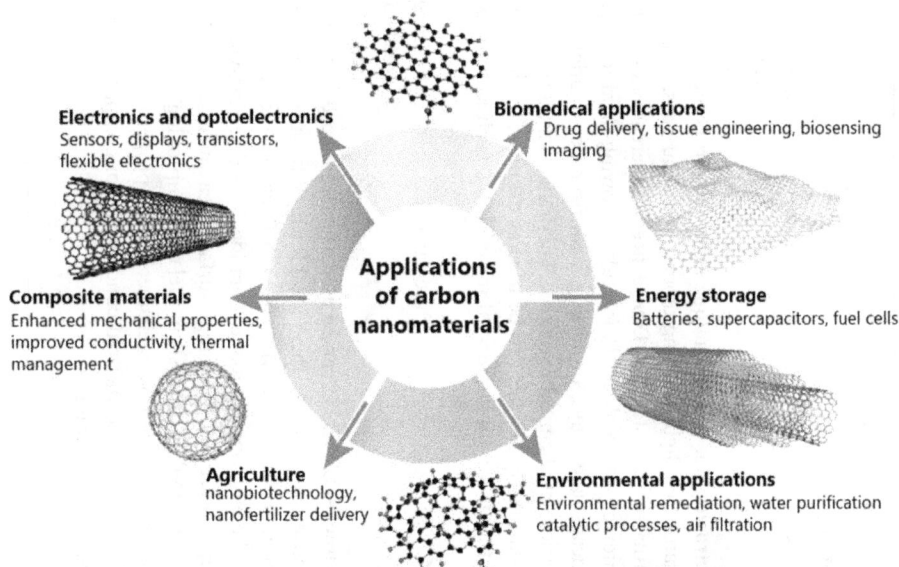

Electronics and optoelectronics
Sensors, displays, transistors, flexible electronics

Biomedical applications
Drug delivery, tissue engineering, biosensing imaging

Applications of carbon nanomaterials

Composite materials
Enhanced mechanical properties, improved conductivity, thermal management

Energy storage
Batteries, supercapacitors, fuel cells

Agriculture
nanobiotechnology, nanofertilizer delivery

Environmental applications
Environmental remediation, water purification catalytic processes, air filtration

Figure 2.8. Diverse applications of carbon nanomaterials: from electronics to medicine.

Table 2.3. Overview of carbon nanomaterials and their applications.

Carbon nanomaterial	Chemistry	Key intrinsic properties	Structure	Preparation methods
Application				
Fullerenes (e.g. C_{60})	Pure carbon	Spherical structure, high stability, unique electronic and optical properties	0D	Arc discharge, laser ablation, chemical vapor deposition

Electronics: due to their unique electronic properties, fullerenes have been explored for applications in electronic devices such as field-effect transistors, light-emitting diodes, and solar cells. They can also be used as electron acceptors in organic photovoltaic cells, improving their efficiency and stability. **Drug delivery:** fullerenes can be used to deliver drugs to cells and tissues. They can be modified with functional groups that interact with biological molecules, allowing them to target specific cells or tissues. Additionally, the hollow structure of fullerenes can be used to encapsulate drugs, protecting them from degradation and improving their bioavailability. **Catalysis:** fullerenes can be used as catalysts for a variety of chemical reactions. They can be modified with functional groups to enhance their catalytic activity or used as supports for other catalysts. Fullerenes can be used as catalysts for reactions such as hydrogenation, oxidation, and polymerization. They can also be used as catalysts for environmental remediation, such as the removal of pollutants from water and air.

SWCNTs	Pure carbon	Cylindrical structure, high mechanical strength, excellent electrical conductivity, high surface area	1D	Arc discharge, laser ablation, chemical vapor deposition

Electronics: SWCNTs can be used as interconnects in electronic devices, such as transistors and integrated circuits. Their high electrical conductivity and small size make them ideal for high-frequency applications. They can also be used as field-effect transistors, which are the building blocks of modern electronic devices. SWCNTs can be used in sensors such as gas sensors and biosensors. Their high surface area and sensitivity to changes in the environment make them ideal for detecting various analytes. **Energy storage:** SWCNTs can be used as electrodes in batteries and supercapacitors. Their high surface area and excellent electrical conductivity make them ideal for storing and releasing electrical energy. They can also be used as catalysts in fuel cells, improving the efficiency of the electrochemical reactions. **Composites:** SWCNTs can be added to composite materials to improve their mechanical properties, such as strength, stiffness, and toughness. They can also be used to improve the electrical conductivity and thermal conductivity of composite materials. SWCNTs can be used in a variety of composite materials, including polymers, metals, and ceramics.

(Continued)

Table 2.3. (*Continued*)

Carbon nanomaterial	Chemistry	Key intrinsic properties	Structure	Preparation methods
MWCNTs	Pure carbon	Cylindrical structure, high mechanical strength, high surface area, excellent thermal conductivity	1D	Arc discharge, laser ablation, chemical vapor deposition
Graphene	Pure carbon	2D sheet, high mechanical strength, excellent electrical and thermal conductivity, large surface area	2D	Mechanical exfoliation, chemical vapor deposition, and epitaxial growth

Composites: MWCNTs can be incorporated into composite materials to enhance their mechanical properties, such as strength, stiffness, and toughness. They can also improve the electrical conductivity and thermal conductivity of composite materials. MWCNTs are particularly suitable for applications in the aerospace, automotive, and construction industries, where lightweight and high-performance materials are required. **Energy storage**: MWCNTs can be used as electrodes in batteries and supercapacitors, where their high surface area and electrical conductivity contribute to improved energy density and power density. They can also be used as supports for catalysts in fuel cells, enhancing the efficiency of the electrochemical reactions. **Sensors**: MWCNTs can be used in various types of sensors, including gas sensors, biosensors, and pressure sensors. Their high surface area and sensitivity to changes in the environment make them ideal for detecting a wide range of analytes. **Environmental remediation**: MWCNTs can be used to remove pollutants from water and air. Their high surface area and adsorption capabilities allow them to efficiently adsorb contaminants such as heavy metals, organic compounds, and dyes. They can also be used as catalysts for the degradation of pollutants, accelerating their removal from the environment.

Electronics: graphene's exceptional electrical conductivity, high electron mobility, and mechanical strength make it a promising material for various electronic applications. It can be used in transistors, integrated circuits, and other electronic devices to improve their performance and efficiency. Graphene can also be used as a transparent conductive electrode in touch screens, solar cells, and other optoelectronic devices. Its high transparency and electrical conductivity make it an ideal material for these applications. **Energy storage**: graphene can be used as an electrode material in batteries and supercapacitors. Its high surface area and electrical conductivity contribute to improved energy density and power density. Graphene can also be used as a catalyst support in fuel cells, enhancing the efficiency of the electrochemical reactions. **Sensors**: graphene can be used in various types of sensors, including gas sensors, biosensors, and chemical sensors. Its high surface area and sensitivity to changes in the environment make it ideal for detecting a wide range of analytes. Graphene can be used to detect gases such as carbon dioxide, nitrogen dioxide, and volatile organic compounds. It can also be used to detect biomolecules such as DNA, proteins, and glucose. Additionally, graphene can be used as a sensor for chemical pollutants, such as heavy metals and pesticides. **Composites**: graphene can be incorporated into composite materials to improve their mechanical properties, electrical conductivity, and thermal conductivity. It can be used to reinforce polymers, metals, and ceramics, creating materials with enhanced performance in various applications. For example, graphene can be added to polymer composites to improve their electrical conductivity and mechanical strength, making them suitable for applications in electronics and aerospace. It can also be added to metal composites to improve their corrosion resistance and mechanical properties, making them suitable for applications in the automotive and construction industries.

Graphene oxide	Carbon with oxygen functional groups	2D sheet, hydrophilic properties, high surface area	2D	Chemical oxidation of graphite

Composites: graphene oxide can be incorporated into composite materials to improve their mechanical properties, electrical conductivity, and thermal conductivity. It can be used to reinforce polymers, metals, and ceramics, creating materials with enhanced performance in various applications. For example, graphene oxide can be added to polymer composites to improve their electrical conductivity and mechanical strength, making them suitable for applications in electronics and aerospace. It can also be added to metal composites to improve their corrosion resistance and mechanical properties, making them suitable for applications in the automotive and construction industries. **Water treatment**: graphene oxide can be used in water treatment applications due to its high surface area and adsorption capabilities. It can be used to remove contaminants from water, such as heavy metals, organic pollutants, and bacteria. Graphene oxide can be used as a filter material or as an adsorbent, depending on the specific application. **Sensors**: graphene oxide can be used in various types of sensors, including gas sensors, biosensors, and chemical sensors. Its high surface area and sensitivity to changes in the environment make it ideal for detecting a wide range of analytes. Graphene oxide can be used to detect gases such as carbon dioxide, nitrogen dioxide, and volatile organic compounds. It can also be used to detect biomolecules such as DNA, proteins, and glucose. Additionally, graphene oxide can be used as a sensor for chemical pollutants, such as heavy metals and pesticides.

Reduced graphene oxide (rGO)	Carbon with reduced oxygen content	2D sheet, conductive properties, high surface area	2D	Chemical reduction of graphene oxide

Electronics: rGO can be used in various electronic devices, such as transistors, sensors, and energy storage devices. Its high electrical conductivity and large surface area make it a promising material for applications in flexible electronics, energy storage, and optoelectronics. For example, rGO can be used as a transparent conductive electrode in solar cells and touch screens. It can also be used as a channel material in field-effect transistors, where its high electron mobility can improve the device's performance. Additionally, rGO can be used as a sensor material for detecting various analytes, such as gases, biomolecules, and pollutants. **Energy storage**: rGO can be used as an electrode material in batteries and supercapacitors. Its high surface area and electrical conductivity contribute to improved energy density and power density. rGO can also be used as a catalyst support for fuel cells, enhancing the efficiency of the electrochemical reactions. **Composites**: rGO can be incorporated into composite materials to improve their mechanical properties, electrical conductivity, and thermal conductivity. It can be used to reinforce polymers, metals, and ceramics, creating materials with enhanced performance in various applications. For example, rGO can be added to polymer composites to improve their electrical conductivity and mechanical strength, making them suitable for applications in electronics and aerospace. It can also be added to metal composites to improve their corrosion resistance and mechanical properties, making them suitable for applications in the automotive and construction industries.

(Continued)

Table 2.3. (*Continued*)

Carbon nanomaterial	Chemistry	Key intrinsic properties	Structure	Preparation methods
Carbon nanofibers	Carbon	Fibrous structure, high mechanical strength, high surface area	1D	Chemical vapor deposition, electrospinning

Composites: carbon nanofibers can be incorporated into composite materials to enhance their mechanical properties, such as strength, stiffness, and toughness. They can also improve the electrical conductivity and thermal conductivity of composite materials. Carbon nanofibers are particularly suitable for applications in the aerospace, automotive, and construction industries, where lightweight and high-performance materials are required. **Energy storage**: carbon nanofibers can be used as electrodes in batteries and supercapacitors, where their high surface area and electrical conductivity contribute to improved energy density and power density. They can also be used as supports for catalysts in fuel cells, enhancing the efficiency of the electrochemical reactions. **Filtration**: carbon nanofibers can be used as filters for various applications, such as water filtration, air filtration, and gas separation. Their high surface area and porous structure allow them to efficiently capture and remove contaminants from fluids and gases. Carbon nanofibers can be used to filter out pollutants, such as heavy metals, organic compounds, and particulate matter, from water and air. They can also be used to separate gases, such as hydrogen and carbon dioxide, from mixtures. Carbon nanofibers can be used in a variety of filtration applications, including drinking water purification, wastewater treatment, and air pollution control.

2.5 Future perspectives and challenges in the field of carbon nanomaterials

Carbon nanomaterials have shown immense potential in various fields, but their widespread adoption and commercialization still face several challenges. Addressing these challenges and exploring future perspectives will be crucial for realizing the full potential of carbon nanomaterials.

Future perspectives:

- **Advancements in synthesis and production:** continued research and development will focus on improving the synthesis and production methods of carbon nanomaterials. This will lead to lower costs, higher quality, and greater scalability, making them more accessible for commercial applications.
- **Novel applications:** as researchers gain a deeper understanding of the properties and behavior of carbon nanomaterials, new and innovative applications will emerge. These could include advancements in electronics, energy storage, healthcare, and environmental remediation.
- **Integration with other materials:** carbon nanomaterials can be combined with other materials to create hybrid materials with enhanced properties. This approach will open up new possibilities for applications in fields such as aerospace, automotive, and construction.
- **Commercialization and market penetration:** efforts will be made to bridge the gap between research and commercialization, ensuring that the benefits of carbon nanomaterials are realized in real-world applications. This will involve addressing regulatory hurdles, intellectual property issues, and supply chain challenges.

Challenges:

- Uniformity and control: achieving consistent quality, uniformity, and control over the properties of carbon nanomaterials remains a significant challenge. This is particularly important for applications that require precise control of their characteristics.
- Scalability and cost: scaling up the production of carbon nanomaterials while maintaining their desired properties and reducing costs is essential for widespread commercialization.
- Toxicity and environmental impact: the potential toxicity and environmental impact of carbon nanomaterials need to be carefully evaluated and addressed. This includes developing safe handling protocols and disposal methods.
- Integration into existing technologies: integrating carbon nanomaterials into existing manufacturing processes and technologies can be challenging. This requires overcoming technical hurdles and ensuring compatibility with existing materials and equipment.
- Intellectual property and commercialization: protecting intellectual property related to carbon nanomaterials and navigating the complex landscape of commercialization can be a significant challenge.

Despite these challenges, the future of carbon nanomaterials looks promising. Continued research and development, coupled with advancements in synthesis, production, and integration, will pave the way for the widespread adoption of these versatile materials.

2.6 Conclusions

Carbon nanomaterials, including fullerenes, carbon nanotubes, and graphene, have emerged as a class of materials with exceptional properties and diverse applications. Their unique structural features, such as their nanoscale dimensions and high surface-to-volume ratios, enable them to exhibit remarkable mechanical, electrical, thermal, and optical properties. This chapter has provided a comprehensive overview of the classification, chemistry, and applications of carbon nanomaterials. The classification of carbon nanomaterials based on their structure and dimensionality has been discussed, highlighting the distinct characteristics of fullerenes, carbon nanotubes, and graphene. The chemistry of carbon nanomaterials, including their surface chemistry, electronic properties, optical properties, and chemical reactivity, has been explored in detail.

Furthermore, the chapter has covered various characterization techniques used to analyze the structure, composition, and properties of carbon nanomaterials. The applications of carbon nanomaterials are vast and span multiple fields. Their exceptional mechanical properties make them suitable for reinforcing composite materials, while their electrical and thermal conductivity makes them promising for electronic and energy storage applications. In the biomedical field, carbon nanomaterials have shown great potential for drug delivery, tissue engineering, and biosensing. Additionally, their high surface area and unique chemical properties make them valuable for environmental remediation and catalytic processes.

Despite the significant advancements made in the field of carbon nanomaterials, several challenges remain. These include the cost-effective production of high-quality carbon nanomaterials, their potential toxicity, and the development of scalable manufacturing processes. Addressing these challenges will be crucial for the widespread adoption of carbon nanomaterials in various industries. As research continues to progress, it is anticipated that carbon nanomaterials will play an even more prominent role in shaping the future of technology and materials science. Their unique properties and versatility offer immense potential for the development of innovative products and solutions across a wide range of applications.

References

[1] Oladipo A, Ogulewe F E, Ansari H, Aleshinloye A O and Gazi M 2024 Carbon Materials as Adsorbents and Catalysts *Advanced Materials for Pharmaceutical Wastewater Treatment* 1st edn (Boca Raton, FL: CRC Press)
[2] Owida H A, Turab N M and Al-Nabulsi J 2023 Carbon nanomaterials advancements for biomedical applications *Bull. Electr. Eng. Informatics.* **12** 891–901

[3]   Zakaria N Z J, Rozali S, Mubarak N M and Ibrahim S 2024 A review of the recent trend in the synthesis of carbon nanomaterials derived from oil palm by-product materials *Biomass Convers. Biorefinery.* **14** 13–44

[4]   Zheng L, Jin W, Xiong K, Zhen H, Li M and Hu Y 2023 Nanomaterial-based biosensors for the detection of foodborne bacteria: a review *Analyst* **148** 5790–804

[5]   Peng Z *et al* 2020 Advances in the application, toxicity and degradation of carbon nanomaterials in environment: a review *Environ. Int.* **134** 105298

[6]   Saleh T A 2020 Nanomaterials: classification, properties, and environmental toxicities *Environ. Technol. Innov.* **20** 101067

[7]   Sajid M 2022 Nanomaterials: types, properties, recent advances, and toxicity concerns *Curr. Opin. Environ. Sci. Heal.* **25** 100319

[8]   de Lara-Castells M P and Mitrushchenkov A O 2021 Mini review: quantum confinement of atomic and molecular clusters in carbon nanotubes *Front. Chem.* **9** 796890

[9]   Gowthaman N S K *et al* 2023 Zero-, one- and two-dimensional carbon nanomaterials as low-cost catalysts in optical and electrochemical sensing of biomolecules and environmental pollutants *Microchem. J.* **194** 109291

[10] Kour R, Arya S, Young S-J, Gupta V, Bandhoria P and Khosla A 2020 Review—Recent advances in carbon nanomaterials as electrochemical biosensors *J. Electrochem. Soc.* **167** 037555

[11] de A C, Mirres M, de B E P, da Silva M, Tessaro L, Galvan D, de Andrade J C, Aquino A, Joshi N and Conte-Junior C A 2022 Recent advances in nanomaterial-based biosensors for pesticide detection in foods *Biosensors* **12** 572

[12] Kuchibhatla S V N T, Karakoti A S, Bera D and Seal S 2007 One-dimensional nano-structured materials *Prog. Mater Sci.* **52** 699–913

[13] Thangaraj B, Solomon P R and Hassan J 2023 Nanocarbon in sodium-ion batteries—a review. Part 2: one, two, and three-dimensional nanocarbons *ChemBioEng Rev.* **10** 647–69

[14] Machín A, Fontánez K, Arango J C, Ortiz D, De León J, Pinilla S, Nicolosi V, Petrescu F I, Morant C and Márquez F 2021 One-dimensional (1D) nanostructured materials for energy applications *Materials (Basel)* **14** 2609

[15] Wu Y H, Yu T and Shen Z X 2010 Two-dimensional carbon nanostructures: Fundamental properties, synthesis, characterization, and potential applications *J. Appl. Phys.* **108** 071301

[16] Jariwala D, Sangwan V K, Lauhon L J, Marks T J and Hersam M C 2013 Carbon nanomaterials for electronics, optoelectronics, photovoltaics, and sensing *Chem. Soc. Rev.* **42** 2824–60

[17] Speranza G 2021 Carbon nanomaterials: synthesis, functionalization and sensing applications *Nanomaterials* **11** 967

[18] Kröner A and Hirsch T 2020 Current trends in the optical characterization of two-dimensional carbon nanomaterials *Front. Chem.* **7** 927

[19] Rizwan M, Shoukat A, Ayub A, Razzaq B and Tahir M B 2021 Types and classification of nanomaterials *Nanomaterials: Synthesis, Characterization, Hazards, and Safety* **2021** (Amsterdam: Elsevier) 31–54

[20] Saha S, Bansal S and Khanuja M 2022 Classification of nanomaterials and their physical and chemical nature *Nano-Enabled Agrochemicals in Agriculture.* (New York, NY: Academic Press) 7–34

[21] Yadav J 2017 Fullerene: properties, synthesis and application *Res. Rev. J. Phys.* **6** 1–6

[22] Panda J, Patra T, Panda P K, Sahu R, Tripathy B C and Biswal A 2024 Fullerenes: Synthesis and Applications *New Forms of Carbon* (Palm Bay, FL: Apple Academic Press) 4 93–122

[23] Kerimbekov D S, Akhanova N E, Gabdullin M T, Abdullin K A, Batryshev D G, Zolotarenko A D, Gavrylyuk N A, Zolotarenko O D and Shchur D V 2022 Features of the synthesis of fullerenes and their derivatives *J. Open Syst. Evol. Probl.* **24** 79–90

[24] Mojica M, Alonso J A and Méndez F 2013 Synthesis of fullerenes *J. Phys. Org. Chem.* **26** 526–39

[25] Morii K, Fujikawa C, Kitagawa H, Iwasa Y, Mitani T and Suzuki T 1997 Electronic properties of fullerenes *Mol. Cryst. Liq. Cryst. Sci. Technol. Sect. A Mol. Cryst. Liq. Cryst.* **296** 357–64

[26] Sachdeva S, Singh D and Tripathi S K 2020 Optical and electrical properties of fullerene C70 for solar cell applications *Opt. Mater. (Amst).* **101** 109717

[27] Shinar J, Vardeny Z V and Kafafi Z H 2000 *Optical and Electronic Properties of Fullerenes and Fullerene-based Materials* (New York, NY: Marcel Dekker)

[28] Hirsch A 2006 Functionalization of fullerenes and carbon nanotubes *Phys. Status Solidi Basic Res.* **243** 3209–12

[29] Zhang W, Sprafke J K, Ma M, Tsui E Y, Sydlik S A, Rutledge G C and Swager T M 2009 Modular functionalization of carbon nanotubes and fullerenes *J. Am. Chem. Soc.* **131** 8446–54

[30] Jia L, Chen M and Yang S 2020 Functionalization of fullerene materials toward applications in perovskite solar cells *Mater. Chem. Front.* **4** 2256–82

[31] Paukov M, Kramberger C, Begichev I, Kharlamova M and Burdanova M 2023 Functionalized fullerenes and their applications in electrochemistry, solar cells, and nano-electronics *Materials (Basel)* **16** 1276

[32] Xu W, Hao Y, Uhlik F, Shi Z, Slanina Z and Feng L 2013 Structural and electrochemical studies of Sm@D3h-C74 reveal a weak metal-cage interaction and a small band gap species *Nanoscale* **5** 10409–13

[33] Lu X, Slanina Z, Akasaka T, Tsuchiya T, Mizorogi N and Nagase S 2010 Yb@C2n ($n = 40$, 41, 42): New fullerene allotropes with unexplored electrochemical properties *J. Am. Chem. Soc.* **132** 5896–905

[34] Zhang Y, Xu J, Hao C, Shi Z and Gu Z 2006 Synthesis, isolation, spectroscopic and electrochemical characterization of some calcium-containing metallofullerenes *Carbon N. Y.* **44** 475–9

[35] Wu B, Hu J, Cui P, Jiang L, Chen Z, Zhang Q, Wang C and Luo Y 2015 Visible-light photoexcited electron dynamics of scandium endohedral metallofullerenes: the cage symmetry and substituent effects *J. Am. Chem. Soc.* **137** 8769–74

[36] Kroto H W, Heath J R, O'Brien S C, Curl R F and Smalley R E 1985 C_{60}: Buckminsterfullerene *Nature* **318** 162–3

[37] Kroto H W, Allaf A W and Balm S P 1991 C_{60}: Buckminsterfullerene *Chem. Rev.* **91** 1213–35

[38] Zhennan G, Jiuxin Q, Xihuang Z, Yongqing W, Xing Z, Sunqi F and Zizhao G 1991 Buckminsterfullerene C60: synthesis, spectroscopic characterization, and structure analysis *J. Phys. Chem.* **95** 9615–8

[39] Liu Q, Cui Q, Li X J and Jin L 2014 The applications of buckminsterfullerene C_{60} and derivatives in orthopaedic research *Connect. Tissue Res.* **55** 71–9

[40] Sariciftci N S 1995 Role of buckminsterfullerene, C_{60}, in organic photoelectric devices *Prog. Quantum Electron.* **19** 131–59

[41] Wudl F and Thompson J D 1992 Buckminsterfullerene C_{60} and organic ferromagnetism *J. Phys. Chem. Solids* **53** 1449–55

[42] Breslavskaya N N, Levin A A and Buchachenko A L 2004 Endofullerenes: size effects on structure and energy *Russ. Chem. Bull.* **53** 18–23

[43] Guha S and Nakamoto K 2005 Electronic structures and spectral properties of endohedral fullerenes *Coord. Chem. Rev.* **249** 1111–32

[44] 2014 *Endohedral Fullerenes: From Fundamentals to Applications* ed S Yang and C R Wang (Singapore: World Scientific)

[45] Wilson L J 1999 Medical applications of fullerenes and metallofullerenes *Electrochem. Soc. Interface* **8** 24

[46] Anzar N, Hasan R, Tyagi M, Yadav N and Narang J 2020 Carbon nanotube—a review on synthesis, properties and plethora of applications in the field of biomedical science *Sensors Int* **1** 100003

[47] Aqel A, El-Nour K M M A, Ammar R A A and Al-Warthan A 2012 Carbon nanotubes, science and technology part (I) structure, synthesis and characterisation *Arab. J. Chem.* **5** 1–23

[48] Huang X, Huang D, Chen J Y, Ye R, Lin Q and Chen S 2020 Fabrication of novel electrochemical sensor based on bimetallic Ce–Ni-MOF for sensitive detection of bisphenol A *Anal. Bioanal. Chem.* **412** 849–60

[49] Wang R, Chen D, Wang Q, Ying Y, Gao W and Xie L 2020 Recent advances in applications of carbon nanotubes for desalination: a review *Nanomaterials.* **10** 1203

[50] Saliev T 2019 The advances in biomedical applications of carbon nanotubes *C* **5** 29

[51] Shoukat R and Khan M I 2021 Carbon nanotubes: a review on properties, synthesis methods and applications in micro and nanotechnology *Microsyst. Technol.* **27** 4183–92

[52] Morais S 2023 Advances and applications of carbon nanotubes *Nanomaterials* **13** 2674

[53] Yang F, Wang M, Zhang D, Yang J, Zheng M and Li Y 2020 Chirality pure carbon nanotubes: growth, sorting, and characterization *Chem. Rev.* **120** 2693–758

[54] Tang D M *et al* 2021 Semiconductor nanochannels in metallic carbon nanotubes by thermomechanical chirality alteration *Science* **374** 1616–20

[55] Jagannatham M, Chandran P, Sankaran S, Haridoss P, Nayan N and Bakshi S R 2020 Tensile properties of carbon nanotubes reinforced aluminum matrix composites: a review *Carbon N. Y.* **160** 14–44

[56] Niu M, Cui C, Tian R, Zhao Y, Miao L, Hao W, Li J, Sui C, He X and Wang C 2022 Mechanical and thermal properties of carbon nanotubes in carbon nanotube fibers under tension-torsion loading *RSC Adv.* **12** 30085–93

[57] Cao K *et al* 2020 Atomic mechanism of metal crystal nucleus formation in a single-walled carbon nanotube *Nat. Chem.* **12** 921–8

[58] Kharlamova M V and Kramberger C 2021 Applications of filled single-walled carbon nanotubes: progress, challenges, and perspectives *Nanomaterials* **11** 2863

[59] Wang J and Lei T 2022 Enrichment of high-purity large-diameter semiconducting single-walled carbon nanotubes *Nanoscale* **14** 1096–106

[60] Liu P *et al* 2023 Direct synthesis of semiconducting single-walled carbon nanotubes toward high-performance electronics *Adv. Electron. Mater.* **9** 2300196

[61] Bhardwaj A K, Gupta S and Raj B 2020 Investigation of parameters for Schottky Barrier (SB) height for schottky barrier based carbon nanotube field effect transistor device *J. Nanoelectron. Optoelectron.* **15** 783–91

[62] Zorn N F and Zaumseil J 2021 Charge transport in semiconducting carbon nanotube networks *Appl. Phys. Rev.* **8** 041318

[63] Arunkumar T, Karthikeyan R, Ram Subramani R, Viswanathan K and Anish M 2020 Synthesis and characterisation of multi-walled carbon nanotubes (MWCNTs) *Int. J. Ambient Energy* **41** 452–6

[64] Rathinavel S, Priyadharshini K and Panda D 2021 A review on carbon nanotube: An overview of synthesis, properties, functionalization, characterization, and the application *Mater. Sci. Eng. B* **268** 115095

[65] 2022 Mechanical properties of carbon nanotubes (CNTs): a review *Eurasian J. Sci. Eng.* **8** 54–68

[66] Díez-Pascual A M 2021 Chemical functionalization of carbon nanotubes with polymers: a brief overview *Macromol.* **1** 64–83

[67] Alavi M, Jabari E and Jabbari E 2021 Functionalized carbon-based nanomaterials and quantum dots with antibacterial activity: a review *Expert Rev. Anti. Infect. Ther.* **19** 35–44

[68] Deline A R, Frank B P, Smith C L, Sigmon L R, Wallace A N, Gallagher M J, Goodwin D G, Durkin D P and Howard Fairbrother D 2020 Influence of oxygen-containing functional groups on the environmental properties, transformations, and toxicity of carbon nanotubes *Chem. Rev.* **120** 11651–97

[69] Tiwari S K, Sahoo S, Wang N and Huczko A 2020 Graphene research and their outputs: status and prospect *J. Sci. Adv. Mater. Devices.* **5** 10–29

[70] Razaq A, Bibi F, Zheng X, Papadakis R, Jafri S H M and Li H 2022 Review on graphene-, graphene oxide-, reduced graphene oxide-based flexible composites: from fabrication to applications *Materials (Basel)* **15** 1012

[71] Dideikin A T and Vul A Y 2019 Graphene oxide and derivatives: the place in graphene family *Front. Phys.* **6** 149

[72] Mbayachi V B, Ndayiragije E, Sammani T, Taj S, Mbuta E R and ullah khan A 2021 Graphene synthesis, characterization and its applications: a review *Results Chem.* **3** 100163

[73] Chandel M, Kaur K, Sahu B K, Sharma S, Panneerselvam R and Shanmugam V 2022 Promise of nano-carbon to the next generation sustainable agriculture *Carbon N. Y.* **188** 461–81

[74] Ghalkhani M, Khosrowshahi E M and Sohouli E 2021 Carbon nano-onions: synthesis, characterization, and application *Handbook of Carbon-Based Nanomaterials* (Amsterdam: Elsevier) 159–207

[75] Ahlawat J, Masoudi Asil S, Guillama Barroso G, Nurunnabi M and Narayan M 2021 Application of carbon nano onions in the biomedical field: recent advances and challenges *Biomater. Sci.* **9** 626–44

[76] Ghalkhani M and Sohouli E 2022 Synthesis of the decorated carbon nano onions with aminated MCM-41/Fe3O4 NPs: morphology and electrochemical sensing performance for methotrexate analysis *Microporous Mesoporous Mater.* **331** 111658

[77] Najafi A S G and Alizadeh T 2022 One-step hydrothermal synthesis of carbon nano onions anchored on graphene sheets for potential use in electrochemical energy storage *J. Mater. Sci., Mater. Electron.* **33** 7444–62

[78] Sang S, Yang S, Guo A, Gao X, Wang Y, Zhang C, Cui F and Yang X 2020 Hydrothermal synthesis of carbon nano-onions from citric acid *Chem.—Asian J.* **15** 3428–31

[79] Xiao J, Han J, Zhang C, Ling G, Kang F and Yang Q H 2022 Dimensionality, function and performance of carbon materials in energy storage devices *Adv. Energy Mater.* **12** 2100775

[80] Adeyemi Oladipo A 2024 Cu^0-doped graphitic carbon quantum dots for rapid electro-chemical sensing of glyphosate herbicide in environmental samples *Microchem. J.* **200** 110294

[81] Zhong M, Zhang M and Li X 2022 Carbon nanomaterials and their composites for supercapacitors *Carbon Energy* **4** 950–85

[82] Mishra S, Mishra S, Patel S S, Singh S P, Kumar P, Khan M A, Awasthi H and Singh S 2022 Carbon nanomaterials for the detection of pesticide residues in food: a review *Environ. Pollut.* **310** 119804

[83] Thakur V K and Thakur M K 2015 *Chemical Functionalization of Carbon Nanomaterials: Chemistry and Applications* (Boca Raton, FL: CRC Press)

[84] Zhu J, Huang Y, Mei W, Zhao C, Zhang C, Zhang J, Amiinu I S and Mu S 2019 Effects of intrinsic pentagon defects on electrochemical reactivity of carbon nanomaterials *Angew. Chem.—Int. Ed.* **58** 3859–64

[85] Rudolph P 2016 Fundamentals and engineering of defects *Prog. Cryst. Growth Charact. Mater.* **62** 89–110

[86] Ngome Okello O F, Yang D H, Chu Y S, Yang S and Choi S Y 2021 Atomic-level defect modulation and characterization methods in 2D materials *APL Mater.* **9** 100902

[87] Olabi A G, Abdelkareem M A, Wilberforce T and Sayed E T 2021 Application of graphene in energy storage device—a review *Renew. Sustain. Energy Rev.* **135** 110026

[88] Bai L, Zhang Y, Tong W, Sun L, Huang H, An Q, Tian N and Chu P K 2020 Graphene for energy storage and conversion: synthesis and interdisciplinary applications *Electrochem. Energy Rev.* **3** 395–430

[89] Chaves A *et al* 2020 Bandgap engineering of two-dimensional semiconductor materials *Npj 2D Mater. Appl.* **4** 29

[90] Hao Q, Li P, Liu J, Huang J and Zhang W 2023 Bandgap engineering of high mobility two-dimensional semiconductors toward optoelectronic devices *J. Mater.* **9** 527–40

[91] Su D, Li H, Yan X, Lin Y and Lu G 2021 Biosensors based on fluorescence carbon nanomaterials for detection of pesticides *TrAC—Trends Anal. Chem.* **134** 116126

[92] Brindhadevi K, AL Garalleh H, Alalawi A, Al-Sarayreh E and Pugazhendhi A 2023 Carbon nanomaterials: types, synthesis strategies and their application as drug delivery system for cancer therapy *Biochem. Eng. J.* **192** 108828

[93] Sikiru S, Oladosu T L, Kolawole S Y, Mubarak L A, Soleimani H, Afolabi L O and Oluwafunke Toyin A O 2023 Advance and prospect of carbon quantum dots synthesis for energy conversion and storage application: a comprehensive review *J. Energy Storage.* **60** 106556

[94] Rasal A S, Yadav S, Yadav A, Kashale A A, Manjunatha S T, Altaee A and Chang J Y 2021 Carbon quantum dots for energy applications: a review *ACS Appl. Nano Mater.* **4** 6515–41

IOP Publishing

Sustainable Carbon Nanomaterials and their Applications

Rafik Naccache and Adedapo O. Adeola

Chapter 3

Scalable synthesis and applications of carbon dots in sustainable technologies

Mahnoor Hassan, Kassa Belay Ibrahim, Tofik Ahmed Shifa, Elisa Moretti and Alberto Vomiero

Carbon dots (C-dots) have emerged as a highly versatile class of carbon-based nanomaterials with remarkable optical, electronic, and chemical properties. Their scalability, tunable photoluminescence, low toxicity, and excellent biocompatibility make them ideal candidates for a wide range of sustainable technologies, including energy conversion, environmental remediation, sensing, and bioimaging. This chapter explores recent advancements in the scalable synthesis of C-dots, high-lighting innovative methods such as microwave-assisted and hydrothermal synthesis that improve control over size, shape, and surface chemistry. It also addresses the critical challenges of synthesis reproducibility, photostability, and environmental compatibility, which are essential for their commercial deployment. Furthermore, the chapter examines emerging trends, including hybrid C-dot systems and their application in next-generation fields such as quantum computing and advanced imaging. By evaluating both the current state and future directions, this chapter underscores the transformative potential of C-dots in enabling green, cost-effective, and high-performance solutions for global sustainability challenges.

3.1 Introduction

The increasing global demand for energy, driven by factors such as population growth, climate change, and fluctuating fossil fuel prices, necessitates a shift towards renewable alternatives [1]. Solar energy emerges as a promising solution due to its environmental benefits, abundance, and potential for widespread implementation. As the most abundant and inexhaustible energy source on Earth, solar energy can be harnessed to generate electricity and heat, offering a sustainable alternative to fossil fuels, which contribute to greenhouse gas emissions [2]. Solar energy is captured using technologies such as photovoltaic (PV) panels, which convert sunlight directly

doi:10.1088/978-0-7503-6325-9ch3
3-1

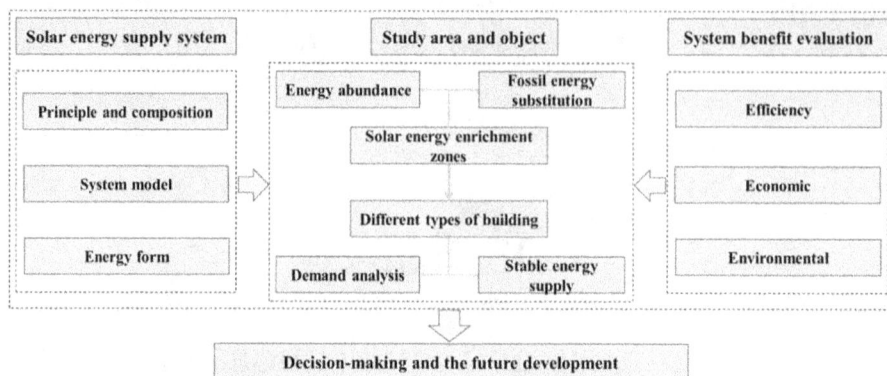

Figure 3.1. Review methodology for energy systems in solar energy enrichment zones. Reprinted from [5], Copyright (2024), with permission from Elsevier.

into electricity. This technology is crucial for powering homes, businesses, and other applications, making solar energy a key component in the transition to a sustainable energy future [3]. Since solar energy does not emit climate-altering gases or other pollutants, it is considered a clean and green source. However, challenges remain, such as the high costs of the technology and the variability of sunlight with location, season, and time of day [4]. Figure 3.1 shows a review methodology that will help in decision-making and future development.

Despite these challenges, solar energy is increasingly being used around the world, particularly in areas with significant quantities of solar radiation, such as the Middle East, Australia, and the southern United States [6–8]. Solar energy is anticipated to become an increasingly significant source of renewable energy in the future as technology advances and costs fall. In many regions of the world, sunlight provides a free and plentiful energy source that can be harvested and converted into electricity. According to some estimates, sunlight that reaches the surface of the Earth contains 10 000 times more energy than the energy currently used [9]. Therefore, it is anticipated that solar energy utilization will be able to satisfy a substantial amount of the projected power consumption. Solar energy is gaining popularity owing to its many benefits, but to date, the method of transforming solar energy into electricity has not been sufficiently effective or affordable [10]. To address this issue, the European Parliament has mandated that from now on, newly constructed buildings must be nearly zero energy by 2030.

3.2 Overview of C-dots

3.2.1 Definition and basic characteristics

C-dots are a relatively new class of carbon-based nanomaterials that have gained significant attention due to their unique properties and potential applications. Typically, C-dots are small nanoparticles with diameters ranging from two to ten nanometers, characterized by a graphitic carbon core and a surface rich in functional

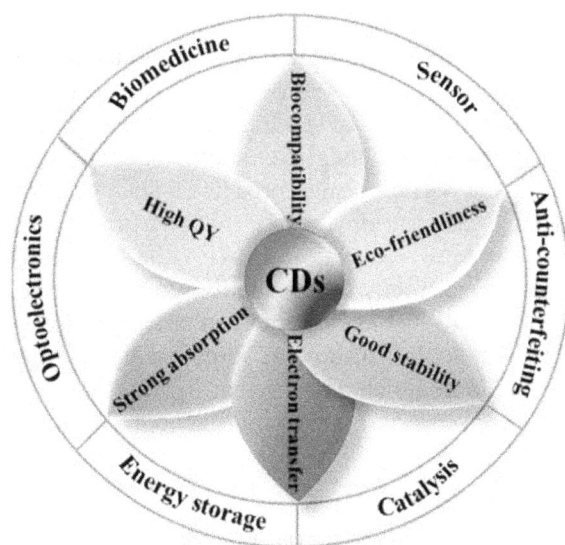

Figure 3.2. Schematic representation of the key properties and applications of C-dots. Reprinted with permission from [13]. Copyright (2020) American Chemical Society.

groups such as carboxyl, hydroxyl, and amino groups [11]. One of the most remarkable properties of C-dots is their strong and tunable photoluminescence. The emission of light by C-dots can be finely adjusted by modifying their size, surface passivation, and heteroatom doping, allowing them to emit light across the visible spectrum [12]. The emitted light can be used in various applications, such as luminescent solar concentrators (LSCs), biomedicine, sensors, optoelectronics, etc. as shown in figure 3.2.

3.2.2 Distinction from other carbon nanomaterials

C-dots are distinct from other carbon nanomaterials such as carbon nanotubes (CNTs), graphene, and fullerenes in several critical ways. Unlike CNTs and graphene, which have well-defined and extended crystalline structures, C-dots are typically amorphous or possess a partially crystalline core, contributing to their unique optical properties [14]. The photoluminescence of C-dots, one of their most distinctive features, is less pronounced in CNTs and graphene, making C-dots more suitable for applications requiring strong fluorescence [15]. Additionally, the synthesis of C-dots is generally simpler and more environmentally friendly. For instance, C-dots can be synthesized using microwave-assisted methods, hydrothermal processes, or even through the simple combustion of organic materials, all of which are relatively low-cost and sustainable [16].

3.2.3 Importance of sustainability in nanomaterial research

Sustainability is a critical consideration in nanomaterial research, driven by the need to minimize environmental impact and ensure economic viability. The development of green synthesis methods for C-dots exemplifies the trend toward more sustainable nanotechnology. Green chemistry approaches, such as using renewable resources, reducing energy consumption, and avoiding hazardous chemicals, are increasingly being employed in the synthesis of C-dots. For example, microwave-assisted synthesis and hydrothermal methods are energy-efficient techniques that can produce high-quality C-dots using benign precursors such as citric acid and natural biomass [17]. These methods not only reduce the environmental footprint of C-dot production but also lower the overall cost, making the technology more accessible and scalable. Moreover, the focus on sustainability in C-dot research is not just about reducing negative impacts but also about enhancing the positive contributions of nanotechnology to society. For instance, C-dots can be used in water purification systems to remove pollutants, in energy-efficient lighting and display technologies, and in biomedical applications that improve health outcomes [15].

3.2.4 Overview of the chapter and its scope

This chapter provides a comprehensive overview of C-dots, delving into their funda-mental characteristics, synthesis methods, and distinguishing features compared to other carbon-based nanomaterials. The chapter begins by defining C-dots and discus-sing their basic properties, including their size, structure, and remarkable photo-luminescence. It then explores the unique aspects of C-dots that set them apart from other nanomaterials, such as carbon nanotubes and graphene, highlighting their simpler synthesis processes, superior biocompatibility, and strong fluorescence. The importance of sustainability in the synthesis and application of C-dots is emphasized, showcasing green chemistry approaches and the use of renewable and waste materials as precursors. The chapter aims to provide a thorough understanding of the current state of C-dot research, covering the latest advancements in their synthesis, characterization, and practical applications. By discussing the sustainable aspects of C-dot production and their potential uses in various fields, the chapter sets the stage for future innovations and research directions in the field of carbon-based nanomaterials. This comprehensive overview serves as a foundation for researchers, engineers, and practitioners interested in the development and application of C-dots in sustainable technologies.

3.3 Green and sustainable synthesis methods of C-dots

Green synthesis focuses on designing chemicals that are safer for human health and the environment. It emphasizes the use of safer solvents, less hazardous chemical synthesis methods, renewable feedstocks, and energy-efficient processes, while also adopting inherently safer chemistry practices to prevent accidents [18]. C-dots can be synthesized using either a top-down or bottom-up approach, as illustrated in figure 3.3. A key characteristic of these synthesis methods is that they result in C-dots with excellent optical properties. In the top-down approach, significantly larger carbon-based components such as graphene, graphene nanotubes, carbon soot, graphite, carbon fibers, and oxide sheets are broken down into sp^2-hybridized carbon domains, structures responsible for fluorescence. This process is achieved through methods such as arc discharge, ultrasonic treatment, and acidic, hydrothermal, and solvothermal exfoliations, as well as laser

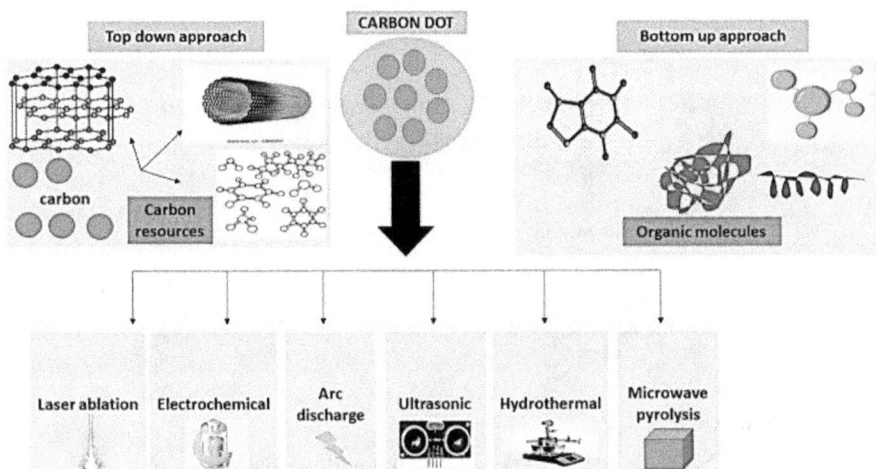

Figure 3.3. Carbon dot synthesis methods. Reprinted from [23], Copyright (2023), with permission from Elsevier. CC BY-NC-ND 4.0

ablation and chemical exfoliation [19, 20]. In the bottom-up methodology, C-dots are synthesized by converting bulk carbon materials into particles through physical and chemical processes such as solvothermal, microwave-assisted, and hydrothermal methods, as well as thermal pyrolysis [19, 21, 22].

3.3.1 Microwave-assisted synthesis

Microwave-assisted synthesis is a rapid and energy-efficient method for producing C-dots. This technique utilizes microwave irradiation to heat the reaction mixture uniformly, leading to fast and controlled synthesis. The high efficiency of microwave heating results in shorter reaction times compared to those of conventional heating methods. Typically, precursors such as citric acid and polyethylene glycol are used, which undergo carbonization under microwave irradiation to form C-dots [24]. This method reduces energy consumption and allows for the use of benign and inexpensive starting materials. Moreover, microwave-assisted synthesis can easily be scaled up, making it suitable for large-scale production [16]. The resulting C-dots often exhibit excellent photoluminescent properties and high quantum yields, making them ideal for applications in bioimaging and optoelectronics [17].

A recent study by Gong *et al* [25] explored the use of a microwave-assisted method to synthesize silicon-doped carbon nanodots (Si-CDs) for use in LSCs, as shown in figure 3.4. In their approach, which applied microwave irradiation to rapidly produce Si-CDs, citric acid and silicon sources were used as precursors. These nanodots exhibited strong visible fluorescence and high photostability, making them effective in enhancing the performance of LSCs by improving light absorption and energy conversion efficiency [25].

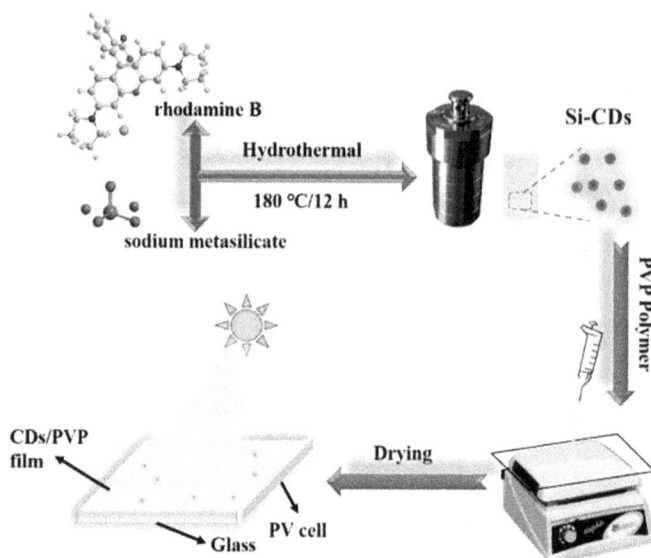

Figure 3.4. Schematic diagrams for the synthesis of Si-CDs and LSC preparation procedure. Reproduced from [25]. CC BY 4.0.

3.3.2 Hydrothermal methods

Hydrothermal synthesis is another green approach for producing C-dots, involving the reaction of carbon precursors in a high-temperature, high-pressure aqueous solution. This method mimics natural geological processes, resulting in well-crystallized nanoparticles with desirable properties. Common precursors include organic compounds such as glucose, citric acid, and plant extracts, which are subjected to hydrothermal treatment to produce C-dots [26]. This thermally mediated approach requires pressurized autoclave vessels, reaction temperatures ranging from 120 °C to 240 °C, and reaction times of 3–12 h for a typical synthesis (table 3.1). The use of water as a solvent aligns with the principles of green chemistry, avoiding the need for toxic organic solvents.

Hydrothermal methods also offer the flexibility to tune the size and surface properties of C-dots by adjusting the reaction parameters, such as temperature, pressure, and reaction time [15]. This technique is particularly advantageous for producing C-dots with high crystallinity and uniform size distribution, which are crucial for applications in sensing and catalysis [14].

A recent study [28] reported the synthesis of highly emissive yellow and red C-dots using a straightforward hydrothermal approach, as shown in figure 3.5. The size of the red C-dots (R-CDs) was approximately 7.02 nm, significantly larger than that of the yellow C-dots (Y-CDs), which measured around 3.45 nm. The quantum yield (QY) of the Y-CDs was found to be 86.4%, while that of the R-CDs was 17.6%. Using these CDs, researchers fabricated a 100 cm^2 tandem LSC with a power conversion efficiency (PCE) of 3.80%, which is among the highest reported in the literature for LSCs of similar size.

Table 3.1. Comparison of various carbon sources and their properties [27].

Carbon source	Method	Conditions	Size [nm]	N: C	O: C	Excitation [nm]	Emission [nm]	QY	References
Water chestnut	Hydrothermal	120 °C, 5 h	3.0	0.05	0.33	380	458	11%	17
Neem tree	Hydrothermal	150 °C, 4 h	3.2	0.09	0.12	340	467	27%	18
Maize	Hydrothermal	180 C, 10 h	3.2	0.02	0.35	350	498	19%	19
Rose-heart radish	Hydrothermal	180 °C, 4 h	3.6	0.13	0.43	330	420	14%	20
Holy basil	Hydrothermal	180 °C, 2 h	4.0	0.15	0.20	350	500	8%	8
Scallion	Hydrothermal	180 °C, 12 h	3.5	0.12	0.27	320	440	3%	21
Strawberry	Hydrothermal	180 °C, 3 h	3.5	0.10	0.34	365	444	6%	22
Pomelo	Hydrothermal	200 °C, 2 h	2.9	0.06	0.35	365	444	7%	23
Willow	Hydrothermal	200 °C, 4 h	4.6	0.10	0.39	350	437	9%	24
Tulsi	Hydrothermal	200 °C, 5 h	5.0	0.09	0.44	360	435	3%	9
Citric acid, glutathione	Hydrothermal	200 °C, 4 h	1.8	0.14	0.68	340	440	75%	24
Citric acid, L-phenylalanine	Hydrothermal	200 °C, 4 h	3.1	0.06	0.35	350	435	23%	7
Citric acid, L-arginine	Hydrothermal	200 °C, 4 h	3.1	0.12	0.46	355	436	26%	6
Ginkgo	Hydrothermal	200 °C, 5 h	11.9	0.08	0.35	350	440	23%	25
Tamarind tree	Hydrothermal	200 °C, 3 h	3.0	0.04	0.42	350	417	10%	26
Green onion	Hydrothermal	200 °C, 3 h	4.2	0.37	0.38	350	413	10%	27
Aibika	Hydrothermal	200 °C, 3 h	3.0	0.08	0.36	330	420	10%	28
Fragrant olive	Hydrothermal	240 °C, 5 h	2.2	0.04	0.38	340	410	19%	19
Walnut	Solvothermal	180 °C, 12 h	1.0	0.10	0.41	400	537	7%	29
Watermelon	Dry heating	220 °C, 2 h	2.0	0.30	0.29	420	440	19%	4
Peanut	Dry heating	220 °C, 3 h	2.2	0.37	0.17	350	440	10%	30
Bamboo	Dry heating	220 °C, 4 h	3.0	0.13	0.35	370	419	5%	5
Glycine, urea	Microwave	800 W, 3 min	3.2	0.40	0.32	320	380	13%	15
Lotus	Microwave	800 W, 6 min	9.4	0.09	0.53	360	435	19%	16

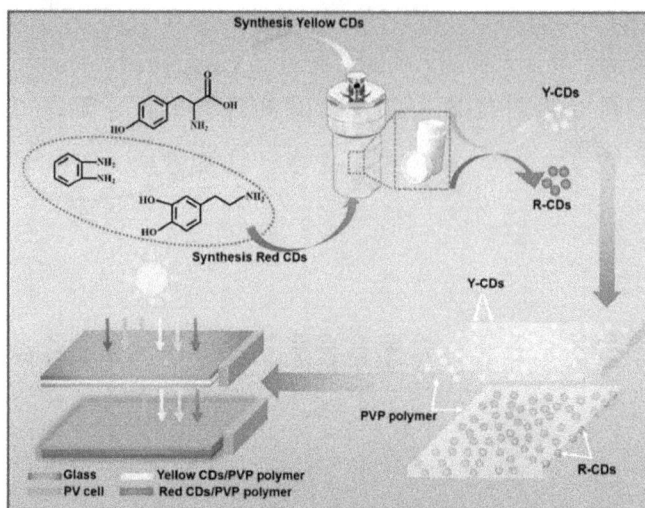

Figure 3.5. Synthesis of C-dots and a schematic of the tandem LSCs based on C-dots. Reproduced from [28] with permission from the Royal Society of Chemistry.

3.3.3 Solvent-free synthesis techniques

Solvent-free synthesis techniques represent the pinnacle of green chemistry and can eliminate the use of solvents, thereby reducing waste and environmental impact. One common approach is the thermal decomposition of organic precursors in a solid-state reaction. For example, citric acid can be thermally decomposed to produce C-dots in the presence of a base such as sodium hydroxide [29]. Solvent-free synthesis (i.e. dry heating) is usually performed at temperatures as high as 300 °C under ambient pressure conditions (table 3.1). These methods are not only environmentally friendly but also cost-effective and utilize readily available and inexpensive materials. Additionally, solvent-free techniques often result in C-dots with unique surface functionalities and excellent luminescent properties, which are beneficial for various applications, including drug delivery and environmental monitoring [30].

Figure 3.6(a) illustrates a practical and green approach to the development of luminescent composites through the in-situ solvent-free formation of C-dots on layered inorganic compounds [31]. Another study highlighted dual-emissive nitrogen-doped carbon dots (N-doped CDs) that were synthesized through the simple, solvent-free annealing of ammonium citrate in air, as depicted in figure 3.6(b) [32].

3.4 Influence of precursor selection on C-dot properties

3.4.1 Impact of different carbon sources

The choice of carbon source plays a crucial role in determining the properties of C-dots. Carbon sources can be broadly categorized into organic compounds, natural products, and waste materials. Organic compounds such as citric acid, glucose, and urea are commonly used due to their high carbon content and ability to yield C-dots with

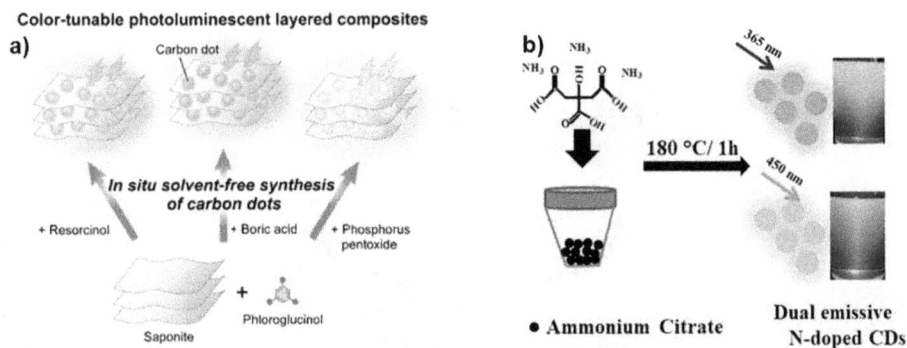

Figure 3.6. Schematic illustration of one-step solvent-free synthesis of carbon-dot-based color-tunable photo-luminescent layered composites (a). Reproduced from [31] with permission from the Royal Society of Chemistry. Graphical overview of the synthesis of dual-emission nitrogen-doped carbon dots (N-doped CDs) using a solvent-free carbonization method (b). Reprinted from [32], Copyright (2022), with permission from Elsevier.

desirable properties [16]. For instance, citric acid is frequently employed because it produces C-dots with strong photoluminescence and high quantum yields. On the other hand, natural products and waste materials, such as fruit peels, biomass, and food residues, offer a sustainable alternative. These sources not only reduce the cost of production but also align with the principles of green chemistry by utilizing renewable and biodegradable materials [26]. The choice of carbon source affects the size, surface functional groups, and optical properties of C-dots. For example, the use of different precursors can result in variations in the emission color and quantum yield of C-dots, which are critical for their applications in sensing, imaging, and electronics [15].

Another example is shown in figure 3.7; multicolor fluorescent C-dots with red (R-CDs), yellow (Y-CDs), and blue (B-CDs) emissions were prepared by choosing proper aromatic precursors with different amounts of benzene rings through a simple solvothermal method. The variation in precursors results in multicolor C-dots that can be used as promising candidates for light-emitting diodes (LEDs) [33].

3.4.2 Role of nitrogen and heteroatom doping for performance enhancement

Doping with nitrogen and other heteroatoms is a well-established strategy for enhancing the performance of C-dots. Nitrogen doping introduces additional electron donors into the carbon lattice, which can improve the electronic properties of C-dots and enhance their photoluminescence [14]. This modification also helps in tuning the emission wavelengths, making it possible to achieve the desired fluorescence colors. In addition to nitrogen, other heteroatoms such as sulfur, phosphorus, and boron can be incorporated into the C-dot structure to further tailor their properties. For example, sulfur and nitrogen doping can enhance the fluorescence intensity and photostability of C-dots [34], while phosphorus doping can improve their optical, sensing, and catalytic properties [35]. The presence of these dopants not only affects the optical properties but also influences the chemical reactivity and stability of C-dots, making them more suitable for

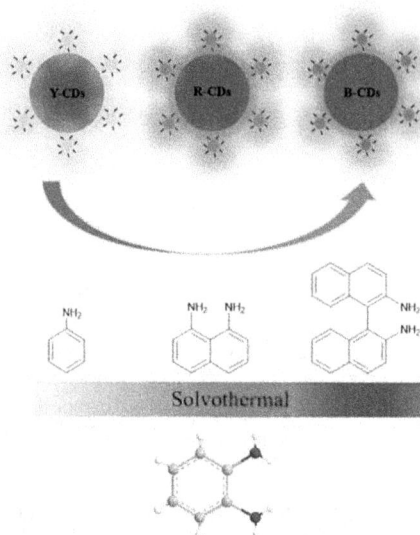

Figure 3.7. Schematic representation of the preparation of C-dots. Reproduced from [33] with permission from the Royal Society of Chemistry.

specific applications such as nanoprobes, optoelectronic devices, catalysis, and bio-medicine [36]. Figure 3.8(a) shows an F-doping strategy that was proposed and adopted to modulate the optical properties of C-dots, whereas figure 3.8(b) illustrates the impact of doping with various materials on electronegativity.

3.4.3 Importance of sustainable solvents

The use of sustainable solvents is crucial in the synthesis of C-dots to minimize environmental impact and promote green chemistry. Traditional synthesis methods often rely on toxic and volatile organic solvents, which can pose health risks and contribute to environmental pollution. In contrast, green chemistry advocates for the use of benign solvents or water, which align with principles of sustainability and safety [39]. For instance, methods that use water or aqueous solutions as solvents are considered more environmentally friendly. These methods avoid the need for hazardous organic solvents and reduce the overall environmental footprint of C-dot production [40]. Some common solvents used in the preparation of C-dots are water, ethanol, dimethylformamide, dimethyl sulfoxide, acetone, toluene, methanol, ethylene glycol, benzene, etc. Figure 3.10 shows some of the effects of using different solvents. In C-dot synthesis, the choice of solvent is crucial, as it directly affects the physicochemical properties of the resulting C-dots, which in turn determines their suitability for various applications. Furthermore, adopting green practices not only helps in reducing waste but also enhances the economic viability

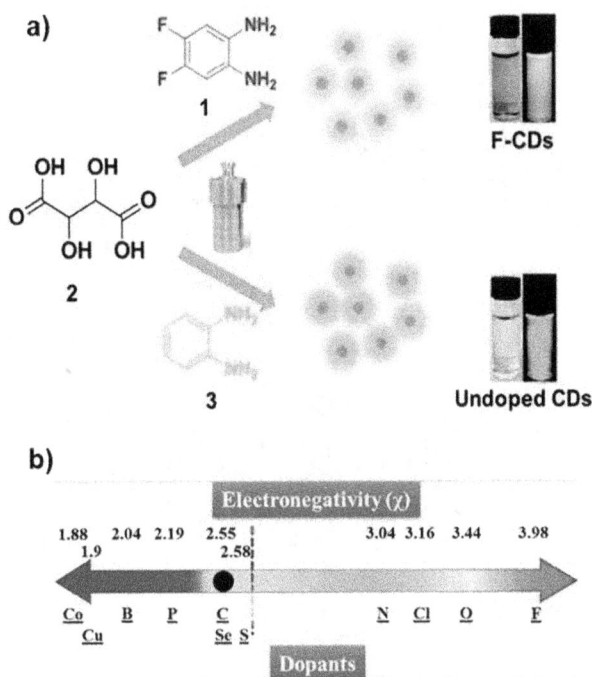

Figure 3.8. Synthesis of F-doped carbon quantum dots (CQDs) and undoped CQDs via the hydrothermal method (a). Reprinted with permission from [37]. Copyright (2017) American Chemical Society. Diagram showing the electronegativity values of different dopants (b). Reproduced from [38] with permission from the Royal Society of Chemistry.

of C-dot production [30]. Figure 3.9 illustrates the effects of different solvents (such as water, DMF, ethanol, and glycerol) on the synthesis and properties of C-dots. It shows how the solvent choice influences the optical properties, particle size, and surface functionalities of the resulting C-dots, highlighting the role of solvent composition and ratios in tailoring C-dot characteristics.

3.5 Tuning optical properties for enhanced performance

3.5.1 Factors influencing fluorescence properties and quantum yield enhancement strategies

The fluorescence properties of C-dots are influenced by several factors, including size, surface functionalization, and the presence of dopants. Quantum yield, a measure of how efficiently C-dots convert absorbed light into emitted fluorescence, can be altered via precursors, fabrication conditions, chemical doping, and surface modifications [46]. For instance, surface passivation with various organic or inorganic molecules can prevent non-radiative recombination and enhance fluorescence efficiency [39]. Additionally, tuning the synthesis conditions, such as

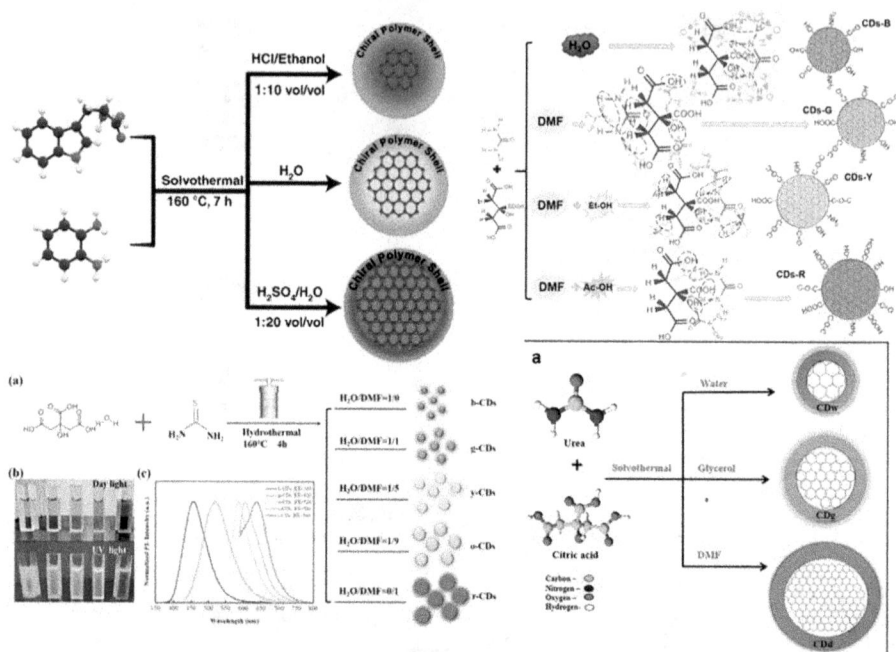

Figure 3.9. Effects of using different solvents in the preparation of C-dots. Top left [41] John Wiley & Sons. © 2021 Wiley-VCH GmbH. Top right reproduced with permission from [42]. © 2021 The Authors. Published by American Chemical Society. CC BY-NC-ND 4.0. Bottom left reproduced from [43]. CC BY 4.0. Bottom right reproduced from [44], with permission from Springer Nature.

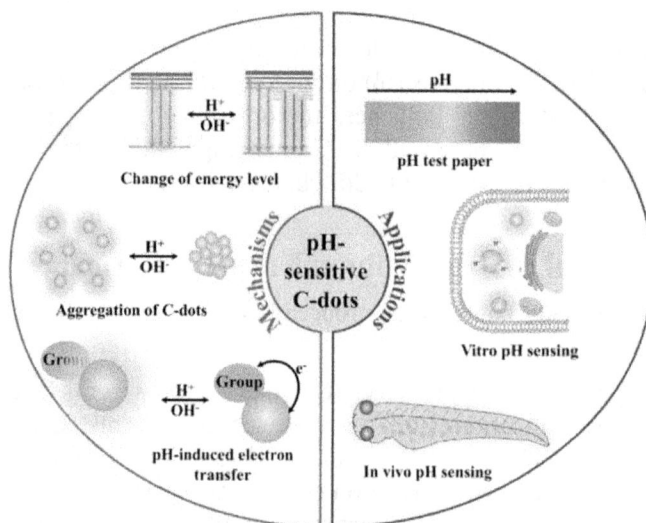

Figure 3.10. Mechanisms and applications of pH-sensitive photoluminescent C-dots. Reproduced from [45]. CC BY 4.0.

temperature and reaction time, can impact the quantum yield by affecting the crystallinity and surface states of C-dots.

3.5.2 Size-dependent emission

The emission properties of C-dots are strongly size-dependent. Generally, smaller C-dots emit higher-energy (blue) fluorescence, while larger C-dots exhibit lower-energy (red) fluorescence. This size-dependent behavior arises from quantum confinement effects, where the energy levels become quantized as the particle size decreases [47]. By controlling the size of C-dots during synthesis, it is possible to tune the emission wavelength across a broad range of the visible spectrum. This ability to adjust the emission color is useful for applications in bioimaging, where different emission wavelengths can be used to label various biomolecules [48].

3.5.3 Surface passivation techniques

Surface passivation is crucial for enhancing the optical properties and stability of C-dots. Surface passivation can form a thin protective layer that insulates C-dots from direct exposure to impurities [49]. Common passivating agents include organic ligands like polyethylene glycol (PEG) or inorganic materials such as silica. Effective passivation reduces surface defects that can act as non-radiative decay centers, thus improving the quantum yield and fluorescence stability [50].

3.5.4 Doping with metals or heteroatoms

Doping C-dots with metals or heteroatoms can significantly alter their optical properties. Metal doping, such as with silver or gold, can introduce localized surface plasmon resonances that enhance fluorescence or change the emission spectrum [51]. Similarly, doping with heteroatoms such as nitrogen, sulfur, or phosphorus can affect the electronic structure and optical properties of C-dots. For example, nitrogen doping can improve photoluminescence and increase the quantum yield by providing additional energy levels within the bandgap [52]. This tuning of properties through doping makes C-dots versatile for various applications, including sensing and catalysis.

3.5.5 Surface functionalization

Surface functionalization involves attaching different chemical groups or molecules to the surface of C-dots to modify their properties. Functionalization can enhance the interaction of C-dots with specific target molecules, improve their stability, and adjust their optical properties. Since there are more reaction sites on the surface of C-dots, they can be functionalized on the surface, allowing carbon quantum dots to have high selectivity, excellent optical properties, and other characteristics, making them widely usable in various fields [53].

3.5.6 Engineering the Stokes shift

The Stokes shift refers to the difference between the absorption and emission peaks of C-dots. Engineering the Stokes shift involves adjusting the photoluminescence properties to achieve a desired separation between these peaks. Techniques used to enhance the Stokes shift include modifying the size of C-dots, changing their surface chemistry, or incorporating specific dopants. LSCs are most effective when they are made from materials that exhibit a large Stokes shift, a high quantum yield, and a broad absorption spectrum [54].

3.5.7 Solvent effect

The solvent chosen for the synthesis and applications of C-dots can substantially impact their optical properties. Solvent polarity and interactions with the C-dots' surface can influence their fluorescence emission and stability [55]. For instance, polar solvents may alter the electronic environment of C-dots, affecting their emission wavelength and intensity. Therefore, selecting an appropriate solvent is crucial for optimizing the performance of C-dots in various applications.

3.5.8 pH and temperature tuning

The optical properties of C-dots can be tuned by adjusting the pH and temperature of their environment. pH changes can affect the surface charge and functional groups of C-dots as shown in figure 3.10, thereby influencing their fluorescence [45]. Temperature also plays a role in the photophysical processes of C-dots; higher temperatures may increase non-radiative relaxation rates, while lower temperatures can stabilize excited states and enhance fluorescence [56]. By carefully controlling these parameters, it is possible to fine-tune the optical properties of C-dots for specific applications.

3.5.9 Size control

Controlling the size of C-dots during synthesis is one of the most effective methods to tune their optical properties. Precise size control can be achieved through various synthesis methods, such as hydrothermal, microwave-assisted, or chemical vapor deposition techniques. Size control allows for the modulation of emission wavelengths and quantum yields, providing flexibility in tailoring C-dots for applications in bioimaging, sensing, and optoelectronics.

3.5.10 Purification of C-dots

Purification of C-dots is crucial for their application in renewable energy, particularly in enhancing the performance of solar cells. In the context of solar cells, the presence of impurities or unreacted precursors in C-dots can negatively affect their ability to absorb sunlight and convert it into electrical energy. Effective purification techniques, such as column chromatography and density gradient centrifugation, help to isolate C-dots with optimal optical properties, which are essential for improving the efficiency of photovoltaic devices. These methods remove non-

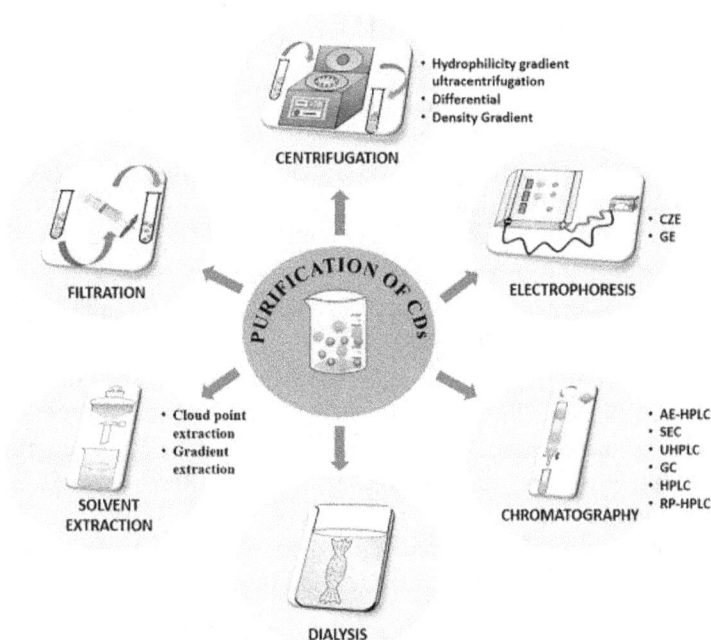

Figure 3.11. Different techniques reported for the purification and separation of CDs. Reproduced from [57]. CC BY 3.0.

fluorescent byproducts and larger aggregates that could otherwise hinder the uniform dispersion of C-dots in the active layer of solar cells. Figure 3.11 illustrates various techniques for the purification of C-dots, including centrifugation, electrophoresis, chromatography, filtration, solvent extraction, and dialysis, each addressing different purification needs to ensure the high purity and quality of the final carbon dot products.

3.6 Structural and surface characterization

3.6.1 Advanced structural analysis techniques

Advanced structural analysis techniques provide detailed insights into the size, shape, and internal structure of C-dots. Techniques such as high-resolution transmission electron microscopy (HRTEM) and scanning transmission electron microscopy (STEM) offer atomic-level resolution, enabling precise determination of the C-dots' crystal lattice and structural defects [58]. Additionally, atomic force microscopy (AFM) provides topographical information about the surface of C-dots, revealing their size distribution and surface roughness [59]. X-ray diffraction (XRD) is another critical technique that helps in identifying the crystalline phases and estimating the size of the graphitic domains within the C-dots [60]. These techniques collectively aid in understanding the structural properties of C-dots, which are crucial for tailoring their optical and electronic behaviors.

3.6.2 Transmission electron microscopy (TEM)

TEM is a powerful tool for characterizing the morphology and size of C-dots at high resolution. TEM allows for direct imaging of C-dots, providing detailed information on their size, shape, and internal structure [61]. HRTEM can further resolve the crystal lattice fringes of the C-dots, revealing the degree of graphitization and the presence of any structural defects [58]. TEM analysis is essential for verifying the uniformity and quality of C-dots, which directly influences their optical properties and performance in various applications.

3.6.3 X-ray photoelectron spectroscopy (XPS)

XPS is employed to analyze the surface chemistry and elemental composition of C-dots. XPS provides information on the elemental composition of the surface, the chemical states of the elements present, and the presence of any functional groups [62]. This technique is beneficial for investigating the effects of surface functional-ization and doping on the chemical environment of C-dots. By analyzing the core-level binding energies, XPS helps understand the interaction of C-dots with other materials and their stability under various conditions.

3.6.4 Spectroscopic techniques (UV–Vis, FTIR, and Raman)

Spectroscopic techniques such as ultraviolet–visible (UV–Vis) spectroscopy, Fourier transform infrared (FTIR) spectroscopy, and Raman spectroscopy are crucial for characterizing the optical and vibrational properties of C-dots.

- **UV–Vis spectroscopy:** this technique is used to study the absorption spectra of C-dots, which provides information on their electronic transitions and optical bandgap. UV–Vis spectroscopy is essential for determining the size distribu-tion and optical behavior of C-dots [63].
- **FTIR spectroscopy:** FTIR spectroscopy helps identify the functional groups on the surface of C-dots by measuring the vibrational modes of molecular bonds. This information is crucial for understanding the surface chemistry and interactions of C-dots [62].
- **Raman spectroscopy:** Raman spectroscopy provides insights into the vibra-tional modes of C-dots and is used to analyze the degree of graphitization and the structural integrity of C-dots. The Raman spectra of C-dots reveal the presence of characteristic peaks associated with the D and G bands, which are indicative of the graphitic structure and defect levels [64].

3.7 Applications in sustainable technologies

3.7.1 Energy harvesting and storage

C-dots are increasingly being explored for energy harvesting and storage applica-tions due to their photoluminescent properties and ability to act as efficient energy converters. In photovoltaic systems, C-dots can enhance the efficiency of solar cells by acting as sensitizers or energy transfer agents, improving the overall light absorption and electron transfer processes [65]. Additionally, C-dots are being investigated for use in supercapacitors and batteries. Their high surface area and excellent electrical conductivity contribute to improved charge storage and cycling stability [66]. The integration of C-dots into energy storage devices could lead to more efficient and sustainable energy solutions.

3.7.2 Solar cells and photovoltaics

In the realm of solar cells and photovoltaics, C-dots offer significant potential for improving the performance of conventional solar technologies. Their ability to absorb and emit light across a broad spectrum allows them to enhance the light absorption in solar cells [67]. C-dots can be incorporated into the active layers of solar cells or used as light-harvesting materials in combination with other semi-conductors. They also serve as excellent electron donors and acceptors, which can help in the creation of more efficient hybrid solar cell systems [68]. Research is ongoing to optimize the integration of C-dots into various types of solar cells, including organic, perovskite, and dye-sensitized solar cells.

3.7.3 Luminescent solar concentrators

LSCs utilize luminescent materials to absorb sunlight and re-emit it at a longer wavelength toward a smaller, high-efficiency photovoltaic cell. C-dots are well suited for this application due to their strong fluorescence and tunable emission properties [69]. By incorporating C-dots into LSCs, researchers aim to improve the efficiency of solar energy conversion and make solar power more accessible and cost-effective. The ability to tailor the emission spectrum of C-dots allows for better matching with the absorption spectra of photovoltaic cells, enhancing the overall performance of LSC systems [25].

3.7.4 Drug delivery and therapy

In the field of drug delivery and therapy, C-dots offer advantages such as biocompatibility, low toxicity, and ease of surface modification. They can be functionalized with various targeting ligands to selectively deliver therapeutic agents to specific cells or tissues [70]. Additionally, C-dots can be used as imaging agents due to their strong fluorescence, allowing for real-time monitoring of drug delivery and therapeutic effects. The high surface area of C-dots also enables the loading of a significant amount of drugs or contrast agents, making them effective for both targeted therapy and diagnostic applications [71].

3.7.5 Water purification

C-dots are being investigated for their potential in water purification applications. Their ability to act as photocatalysts under UV or visible light can facilitate the degradation of organic pollutants and pathogens in contaminated water [72]. By incorporating C-dots into filtration systems and photocatalytic reactors, researchers aim to develop more efficient and sustainable methods for water treatment. The high

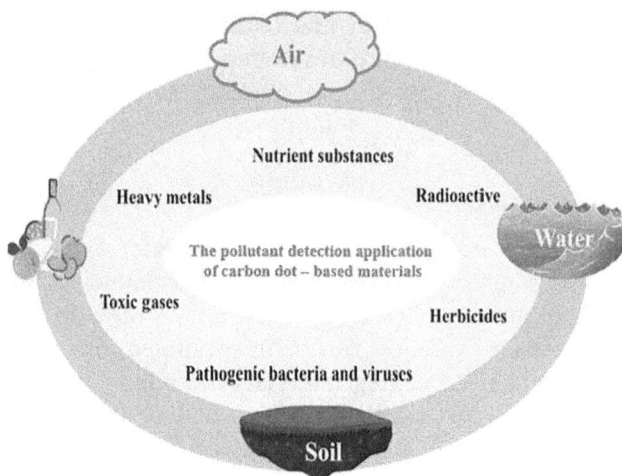

Figure 3.12. Pollutant detection applications of C-dots. Reprinted from [75], Copyright (2023), with permission from Elsevier.

surface area and tunable optical properties of C-dots contribute to their effectiveness in breaking down harmful substances and improving water quality [73].

3.7.6 Sensor and environmental monitoring

C-dots are also utilized in sensors and environmental monitoring due to their fluorescence properties and sensitivity to environmental changes. They can be used to detect various pollutants, such as heavy metals and toxic gases, through fluorescence quenching or enhancement mechanisms [74]. Figure 3.12 shows the application of C-dots in pollution detection. The ability to functionalize C-dots with specific receptors or ligands allows for the selective detection of target analytes in complex environmental samples. This makes C-dots valuable tools for real-time monitoring and environmental protection [75].

3.8 Challenges and future perspectives

3.8.1 Current challenges in optimizing C-dots for enhanced performance

One of the primary challenges in optimizing C-dots is achieving consistent and reproducible synthesis. The synthesis methods often lead to variations in size, shape, and surface chemistry, which can affect the optical and electronic properties of the C-dots [76]. Another challenge is the scalability of synthesis processes, which must be efficient and cost-effective for commercial applications. Additionally, the stability and photobleaching of C-dots under prolonged exposure to light or environmental conditions remain significant issues. Ensuring that C-dots maintain their performance and functionality over time is crucial for their practical use. Furthermore, the development of scalable and environmentally friendly synthesis methods is essential to enhance the sustainability of C-dot production.

3.8.2 Emerging trends and future directions in the field

Recent advancements in C-dot research are focusing on overcoming these challenges and exploring new applications. Emerging trends include the development of novel synthesis techniques that offer better control over the size and surface properties of C-dots, such as microwave-assisted synthesis and hydrothermal methods [19]. Researchers are also investigating the integration of C-dots with other nanomaterials and polymers to create hybrid materials with enhanced properties for specific applications. Another exciting direction is the exploration of C-dots in new areas such as quantum computing and advanced imaging techniques [50, 77]. The future of C-dot research lies in addressing current limitations, expanding their applications, and developing sustainable production methods.

3.9 Conclusions

In conclusion, C-dots represent a versatile and promising material with a wide range of applications in various fields. While significant progress has been made, challenges such as synthesis consistency, scalability, and stability need to be addressed to fully realize their potential. By focusing on innovative synthesis methods and exploring new applications, the field of C-dots is poised for substantial growth. Continued research and development will be crucial in advancing the performance and applicability of C-dots, paving the way for their integration into cutting-edge technologies and sustainable solutions [78].

References

[1] Hunt L C and Kipouros P 2023 Energy demand and energy efficiency in developing countries *Energies* **16** 1056
[2] Kuşkaya S, Bilgili F, Mugǎloǧlu E, Khan K, Hoque M E and Toguç N 2023 The role of solar energy usage in environmental sustainability: fresh evidence through time-frequency analyses *Renew. Energy* **206** 858–71
[3] Hasan M M, Hossain S, Mofijur M, Kabir Z, Badruddin I A, Yunus Khan T M and Jassim E 2023 Harnessing solar power: a review of photovoltaic innovations, solar thermal systems, and the dawn of energy storage solutions *Energies* **16** 6456
[4] Rodríguez F, Fleetwood A, Galarza A and Fontán L 2018 Predicting solar energy generation through artificial neural networks using weather forecasts for microgrid control *Renew. Energy* **126** 855–64
[5] Wang B, Liu Y, Wang D, Song C, Fu Z and Zhang C 2024 A review of the photothermal-photovoltaic energy supply system for building in solar energy enrichment zones *Renew. Sustain. Energy Rev.* **191** 114100
[6] Hassan Q, Al-Hitmi M, Tabar V S, Sameen A Z, Salman H M and Jaszczur M 2023 Middle East energy consumption and potential renewable sources: an overview *Clean. Eng. Technol.* **12** 100599
[7] Tabassum S, Rahman T, Islam A U, Rahman S, Dipta D R, Roy S, Mohammad N, Nawar N and Hossain E 2021 Solar energy in the United States: development, challenges and future prospects *Energies* **14** 8142

[8] Narayanan R, Parthkumar P and Pippia R 2021 Solar energy utilisation in Australian homes: a case study *Case Stud. Therm. Eng.* **28** 101603

[9] Martinez-Gracia A, Arauzo I and Uche J 2019 Solar energy availability *Solar Hydrogen Production* (New York: Academic) 113–49

[10] Vodapally S N and Ali M H 2022 A comprehensive review of solar photovoltaic (PV) technologies, architecture, and its applications to improved efficiency *Energies* **16** 319

[11] Tarannum N, Pooja K, Singh M and Panwar A 2024 A study on the development of C-dots via green chemistry: a state-of-the-art review *Carbon Lett.* **34** 1537–68

[12] Mikhail M M, Ahmed H B, Abdallah A E, El-Shahat M and Emam H E 2024 Surface passivation of carbon dots for tunable biological performance *J. Fluoresc.* **35** 4225–42

[13] Liu J, Li R and Yang B 2020 Carbon dots: a new type of carbon-based nanomaterial with wide applications *ACS Cent. Sci.* **6** 2179–95

[14] Wang X, Cao L, Lu F, Meziani M J, Li H, Qi G, Zhou B, Harruff B A, Kermar- rec F and Sun Y P 2009 Photoinduced electron transfers with carbon dots *Chem. Commun.* **2009** 3774–6

[15] Lim S Y, Shen W and Gao Z 2015 Carbon quantum dots and their applications *Chem. Soc. Rev.* **44** 362–81

[16] Baker S N and Baker G A 2010 Luminescent carbon nanodots: emergent nanolights *Angew. Chem. Int. Ed.* **49** 6726–44

[17] Ghosh A *et al* 2024 Single-step low-temperature synthesis of carbon dots for advanced multiparametric bioimaging probe applications *ACS Appl. Bio Mater.* **7** 7895–908

[18] Anastas P T and Warner J C 2000 *Green Chemistry: Theory and Practice* (Oxford: Oxford University Press)

[19] Namdari P, Negahdari B and Eatemadi A 2017 Synthesis, properties and biomedical applications of carbon-based quantum dots: an updated review *Biomed. pharmacother.* **87** 209–22

[20] Newman Monday Y, Abdullah J, Yusof N A, Abdul Rashid S and Shueb R H 2021 Facile hydrothermal and solvothermal synthesis and characterization of nitrogen-doped carbon dots from palm kernel shell precursor *Appl. Sci.* **11** 1630

[21] Xie R, Wang Z, Zhou W, Liu Y, Fan L, Li Y and Li X 2016 Graphene quantum dots as smart probes for biosensing *Anal. Methods* **8** 4001–16

[22] Yuan F, Li S, Fan Z, Meng X, Fan L and Yang S 2016 Shining carbon dots: synthesis and biomedical and optoelectronic applications *Nano Today* **11** 565–86

[23] Manzoor S, Dar A H, Dash K K, Pandey V K, Srivastava S, Bashir I and Khan S A 2023 Carbon dots applications for development of sustainable technologies for food safety: a comprehensive review *Appl. Food Res.* **3** 100263

[24] Zhu S *et al* 2011 Strongly green-photoluminescent graphene quantum dots for bioimaging applications *Chem. Commun.* **47** 6858–60

[25] Gong X, Zheng S, Zhao X and Vomiero A 2022 Engineering high-emissive silicon-doped carbon nanodots towards efficient large-area luminescent solar concentrators *Nano Energy* **101** 107617

[26] Sun Y-P, Zhou B, Lin Y, Wang W, Fernando K A S, Pathak P and Xie S-Y 2006 Quantum-sized carbon dots for bright and colorful photoluminescence *J. Am. Chem. Soc.* **128** 7756–7

[27] Chahal S, Macairan J R, Yousefi N, Tufenkji N and Naccache R 2021 Green syn- thesis of carbon dots and their applications *RSC Adv.* **11** 25354–63

[28] Li J, Zhao H, Zhao X and Gong X 2021 Red and yellow emissive carbon dots integrated tandem luminescent solar concentrators with significantly improved efficiency *Nanoscale* **13** 9561–9

[29] Shi B, Sun W, Liu Q and Lu C 2023 Ultra-long room temperature phosphorescence carbon dots-based composites with high environmental stability *J. Lumin.* **259** 119834

[30] Nandy S, Fortunato E and Martins R 2022 Green economy and waste management: an inevitable plan for materials science *Prog. Nat. Sci.:Mater. Int.* **32** 1–9

[31] Uchida J, Takahashi Y, Katsurao T and Sakabe H 2022 One-step solvent-free synthesis of carbon dot-based layered composites exhibiting color-tunable photoluminescence *RSC Adv.* **12** 8283–9

[32] Khan W U, Zhou P, Qin L, Alam A, Ge Z and Wang Y 2022 Solvent-free synthesis of nitrogen-doped carbon dots with dual emission and their biological and sensing applications *Mater. Today Nano* **18** 100205

[33] Liu Z, Lu X, Liu M and Wang W 2957 2023. Blue, yellow, and red carbon dots from aromatic precursors for light-emitting diodes *Molecules* **28** 2957

[34] Hemmati A, Emadi H and Nabavi S R 2023 Green synthesis of sulfur-and nitrogen-doped carbon quantum dots for determination of L-DOPA using fluorescence spectroscopy and a smartphone-based fluorimeter *ACS Omega* **8** 20987–99

[35] Yashwanth H J, Rondiya S R, Dzade N Y, Dhole S D, Phase D M and Hareesh K 2020 Enhanced photocatalytic activity of N, P, co-doped carbon quantum dots: an insight from experimental and computational approach *Vacuum* **180** 109589

[36] Miao S, Liang K, Zhu J, Yang B, Zhao D and Kong B 2020 Hetero-atom-doped carbon dots: doping strategies, properties and applications *Nano Today* **33** 100879

[37] Zuo G, Xie A, Li J, Su T, Pan X and Dong W 2017 Large emission red-shift of carbon dots by fluorine doping and their applications for red cell imaging and sensitive intracellular Ag + detection *J. Phys. Chem.* C **121** 26558–65

[38] Li L and Dong T 2018 Photoluminescence tuning in carbon dots: surface passivation or/and functionalization, heteroatom doping *J. Mater. Chem.* C **6** 7944–70

[39] Zhou J, Booker C, Li R, Zhou X, Sham T K, Sun X and Ding Z 2007 An elec-electrochemical avenue to blue luminescent nanocrystals from multiwalled carbon nanotubes (MWCNTs) *J. Am. Chem. Soc.* **129** 744–5

[40] Keçili R, Hussain C G and Hussain C M 2023 Fluorescent nanosensors based on green carbon dots (CDs) and molecularly imprinted polymers (MIPs) for environmental pollutants: emerging trends and future prospects *Trends Environ. Anal. Chem.* **40** e00213

[41] Ru Y, Sui L, Song H, Liu X, Tang Z, Zang S Q, Yang B and Lu S 2021 Rational design of multicolor-emitting chiral carbonized polymer dots for full-color and white circularly polarized luminescence *Angew. Chem.* **133** 14210–8

[42] Huo X, Shen H, Liu R and Shao J 2021 Solvent effects on fluorescence properties of carbon dots: implications for multicolor imaging *ACS Omega* **6** 26499–508

[43] Yu R, Liang S, Ru Y, Li L, Wang Z, Chen J and Chen L 2022 A facile preparation of multicolor carbon dots *Nanoscale Res. Lett.* **17** 32

[44] Yan F, Sun Z, Zhang H, Sun X, Jiang Y and Bai Z 2019 The fluorescence mechanism of carbon dots, and methods for tuning their emission color: a review *Microchim. Acta* **186** 1–37

[45] Liu C, Zhang F, Hu J, Gao W and Zhang M 2021 A mini review on pH-sensitive photoluminescence in carbon nanodots *Front. Chem.* **8** 605028

[46] Anpalagan K, Yin H, Cole I, Zhang T and Lai D T 2024 Quantum yield enhancement of carbon quantum dots using chemical-free precursors for sensing Cr (VI) ions *Inorganics* **12** 96

[47] Zhu S, Song Y, Wang J, Wan H, Zhang Y, Ning Y and Yang B 2017 Photo-luminescence mechanism in graphene quantum dots: quantum confinement effect and surface/edge state *Nano Today* **13** 10–4

[48] Karthik C S, Skorjanc T and Shetty D 2024 Fluorescent covalent organic frameworks–promising bioimaging materials *Mater. Horiz.* **11** 2077–94

[49] Singh A, Qu Z, Sharma A, Singh M, Tse B, Ostrikov K, Popat A, Sonar P and Kumeria T 2023 Ultra-bright green carbon dots with excitation-independent fluorescence for bioimaging *J. Nanostructure Chem.* **13** 377–87

[50] Karami M H, Abdouss M, Rahdar A and Pandey S 2024 Graphene quantum dots: background, synthesis methods, and applications as nanocarriers in drug delivery and cancer treatment: an updated review *Inorg. Chem. Commun.* **161** 112032

[51] Emam A N, Mostafa A A, Mohamed M B, Gadallah A S and El-Kemary M 2018 Enhancement of the collective optical properties of plasmonic hybrid carbon dots via localized surface plasmon *J. Lumin.* **200** 287–97

[52] Rodríguez-Carballo G, Moreno-Tost R, Fernandes S, da Silva J C E, da Silva L P, Galiano E C and Algarra M 2023 Nitrogen-doped carbon dots as a photocatalyst based on biomass. A life cycle assessment *J. Clean. Prod.* **423** 138728

[53] Yang H L, Bai L F, Geng Z R, Chen H, Xu L T, Xie Y C, Wang D J, Gu H W and Wang X M 2023 Carbon quantum dots: preparation, optical properties, and biomedical applications *Mater. Today Adv.* **18** 100376

[54] Zhang X, Liu Y, Kuan C H, Tang L, Krueger T D, Yeasmin S, Ullah A, Fang C and Cheng L J 2023 Highly fluorescent nitrogen-doped carbon dots with large Stokes shifts *J. Mater. Chem.* C **11** 11476–85

[55] Lee H J, Jana J, Ngo Y L T, Wang L L, Chung J S and Hur S H 2019 The effect of solvent polarity on emission properties of carbon dots and their uses in colorimetric sensors for water and humidity *Mater. Res. Bull.* **119** 110564

[56] Zhang H, You J, Wang J, Dong X, Guan R and Cao D 2020 Highly luminescent carbon dots as temperature sensors and 'off-on' sensing of Hg^{2+} and biothiols *Dyes Pigm.* **173** 107950

[57] Ullal N, Mehta R and Sunil D 2024 Separation and purification of fluorescent carbon dots–an unmet challenge *Analyst* **149** 1680–700

[58] Xu Q, Liu Y, Gao C, Wei J, Zhou H, Chen Y, Dong C, Sreeprasad T S, Li N and Xia Z 2015 Synthesis, mechanistic investigation, and application of photoluminescent sulfur and nitrogen-centered carbon dots *J. Mater. Chem.* C **3** 9885–93

[59] Ventrella A, Camisasca A, Fontana A and Giordani S 2020 Synthesis of green fluorescent carbon dots from carbon nano-onions and graphene oxide *RSC Adv.* **10** 36404–12

[60] Mewada A, Pandey S, Shinde S, Mishra N, Oza G, Thakur M, Sharon M and Sharon M 2013 Green synthesis of biocompatible carbon dots using aqueous extract of *Trapa bispinosa* peel *Mater. Sci. Eng.* C **33** 2914–7

[61] Javed N and O'Carroll D M 2021 Long-term effects of impurities on the particle size and optical emission of carbon dots *Nanoscale Adv.* **3** 182–9

[62] Cao L, Zan M, Chen F, Kou X, Liu Y, Wang P, Mei Q, Hou Z, Dong W F and Li L 2022 Formation mechanism of carbon dots: from chemical structures to fluorescent behaviors *Carbon* **194** 42–51

[63] Hamid M *et al* 2024 Solvothermal synthesized N–S doped carbon dots derived from Cavendish banana peel (*Musa paradisiaca*) for the detection of Fe (III) and Pb (II) *Case Stud. Chem. Environ. Eng.* **10** 100832

[64] Singh N S and Giri P K 2024 Ultrasmall graphene quantum dots as exceptionally stable and efficient substrates for graphene-enhanced raman spectroscopy detection of dyes *ACS Appl. Nano Mater.* **7** 13579–89

[65] Li W *et al* 2024 Enhanced light-harvesting and energy transfer in carbon dots embedded thylakoids for photonic hybrid capacitor applications *Angew. Chem.* **136** e202308951

[66] Jin Y *et al* 2024 Recent advances of carbon dots-based emerging materials for supercapacitor applications *J. Energy Storage* **85** 1111–18

[67] Rajhi A A, Abd Alaziz K M, Oviedo B S R, Yadav A, Hernańdez E, Gallegos C, Alamri S and Duhduh A A 2024 Enhancing the performance of quantum dot solar cells through halogen adatoms on carboxyl edge-functionalized graphene quantum dots *J. Photochem. Photobiol., A* **447** 115240.

[68] Srivastava I, Khamo J S, Pandit S, Fathi P, Huang X, Cao A, Haasch R T, Nie S, Zhang K and Pan D 2019 Influence of electron acceptor and electron donor on the photophysical properties of carbon dots: a comparative investigation at the bulk-state and single-particle level *Adv. Funct. Mater.* **29** 1902466

[69] Li J, Zhao H, Zhao X and Gong X 2024 High-efficiency luminescent solar concentrators based on carbon dots with simultaneously ultrabright solid-state and liquid-state luminescence *Adv. Funct. Mater.* **34** 2404473

[70] Medina-Berríos N, Pantoja-Romero W, Lavín Flores A, Díaz Vélez S C, Martínez Guadalupe A C, Torres Mulero M T, Kisslinger K, Martínez-Ferrer M, Morell G and Weiner B R 2024 Synthesis and characterization of carbon-based quantum dots and doped derivatives for improved andrographolide's hydrophilicity in drug delivery platforms *ACS Omega* **9** 12575–84

[71] Wang J, Fu Y, Gu Z, Pan H, Zhou P, Gan Q, Yuan Y and Liu C 2024 Multifunctional carbon dots for biomedical applications: diagnosis, therapy, and theranostics *Small* **20** 2303773

[72] Parambil A M, Rajan S, Huang P C, Shashikumar U, Tsai P C, Rajamani P, Lin Y C and Ponnusamy V K 2024 Carbon and graphene quantum dots-based architectonics for efficient aqueous decontamination by adsorption chromatography technique-current state and prospects *Environ. Res.* **251** 118541

[73] Singh P, Rani N, Kumar S, Kumar P, Mohan B, Bhankar V, Kataria N, Kumar R and Kumar K 2023 Assessing the biomass-based carbon dots and their composites for photo-catalytic treatment of wastewater *J. Clean. Prod.* **413** 137474

[74] Hebbar A, Selvaraj R, Vinayagam R, Varadavenkatesan T, Kumar P S, Duc P A and Rangasamy G 2023 A critical review of the environmental applications of carbon dots *Chemosphere* **313** 137308

[75] Tran N A, Hien N T, Hoang N M, Dang H L T, Van Quy T, Hanh N T, Vu N H and Dao V D 2023 Carbon dots in environmental treatment and protection applications *Desalination* **548** 116285

[76] Liu H, Zhong X, Pan Q, Zhang Y, Deng W, Zou G, Hou H and Ji X 2024 A review of carbon dots in synthesis strategy *Coord. Chem. Rev.* **498** 215468

[77] Zhu P *et al* 2023 Exploring multi-element co-doped carbon dots as dual-mode probes for fluorescence/CT imaging *Chem. Eng. J.* **470** 144042

[78] Chary K J S, Sharma A and Singh A 2023 Carbon quantum dots in healthcare: a promising solution for sustainable healthcare and biomedical practices *E3S Web of Conferences* **453** (EDP Sciences) 01017

IOP Publishing

Sustainable Carbon Nanomaterials and their Applications

Rafik Naccache and Adedapo O. Adeola

Chapter 4

Revolutionizing carbon nanomaterial synthesis: computational approaches for sustainability

Md Rifat Khandaker and Jahanara Sarker Ayesha

This chapter provides a structured review of how computational tools, such as density functional theory (DFT), molecular dynamics (MD), and machine learning (ML), have led to transformative advancements in sustainable synthesis pathways for carbon nanomaterials (CNMs). Conventional methods for CNM production suffer from numerous environmental issues, including high energy and water use, inefficient resource utilization, and waste disposal. Comparatively, a computational strategy offers a new way of thinking, enabling optimal control of synthesis parameters to yield high-quality CNMs while minimizing environmental impact. The above analysis characterizes recent interdisciplinary advancements in DFT-informed reaction optimization, MD-facilitated process simulation, and ML-guided predictive modeling in materials design. These methods are expounded upon using key use cases such as energy storage, catalysis, and environmental remediation, which serve to fast-track eco-friendly CNM innovation. The chapter closes with future-oriented insights into the expansion of computational approaches to address unresolved areas in sustainable nanomanufacturing, highlighting the essential connection between theoretical insights and experimental verification. This chapter serves as a guide for prospective learners seeking to harness computational abilities to design sustainable nanomaterials and realize common goals for a sustainable future.

4.1 The role of computational tools in sustainable carbon nanomaterial synthesis

The growing interest in green technology is changing the way materials are developed in various sectors. The properties of CNMs make them suitable for a variety of applications, including energy storage, catalysis, biosensing, drug discovery, and environmental monitoring [1–3]. The traditional preparation processes, including chemical vapor deposition, arc discharge, and laser ablation approaches, are facing serious environmental pollution and cost issues. These methods utilize hazardous chemicals, require a lot of energy, and produce waste during fabrication. As previously discussed, these drawbacks raise the question of how CNM synthesis can be carried out sustainably without causing environmental pollution, excessive energy demand, and resource depletion [4]. Computational tools enable the simulation and prediction of various properties and behaviors of CNMs at a precise level without carrying out exhaustive physical experiments. These methods provide an improved understanding of CNM properties and enable the design of new materials [5]. Computational methods, including DFT, MD, and ML, are revolutionizing the sustainable production of CNMs by enabling accurate property prediction and comparative analysis of various synthesis techniques, thereby enhancing both efficiency and sustainability.

4.1.1 Density functional theory

DFT is a quantum mechanical simulation method that can provide an analysis of the electronic properties and stability of CNMs at the atomic scale [6]. At the same time, it provides essential analysis of the electronic structure, reactivity, and thermodynamics of CNMs, and it is very beneficial in materials design. In addition, DFT is necessary to predict electronic behavior, calculate the bandgap, and respond to structural stability in different environments in which CNMs might be used, allowing the properties of CNMs to be tuned and optimized for a specific purpose, for example, energy storage devices or for improved structural strength [7, 8].

4.1.2 Molecular dynamics simulations

MD simulations are effective complementary methods used to obtain a more complete description of CNMs' behavior during their production, growth, and postprocessing [9]. The capability of MD simulations to account for molecular interactions and kinetics complements the DFT contribution to the description of the static properties of materials [10]. MD simulations are useful for deriving more information about synthesis conditions and predicting CNMs' behavior in thermal, pressure, or chemical environments. On the other hand, DFT, as an instrument for acquiring profound knowledge about electronic properties, electronic structure, and reactivity, as well as the thermodynamics of CNMs, has become an essential tool in materials science. DFT, which is used to predict the electronic properties of CNMs and calculate the bandgap values and their structural stability effects, promotes optimal resource utilization and reduces pollution [11].

4.1.3 Machine learning in nanomaterial discovery

ML has been central to new material discovery and design. ML approaches, such as supervised, unsupervised, and reinforcement learning, have been used to analyze extensive experimental and simulated data to extract key patterns and correlations from material data [12–14]. ML enables scientists to engage in high-throughput testing of various candidate carbon nanotubes (CNTs) and provides algorithms that quickly calculate conductivity, surface area, and stability. ML is also vital in calibrating synthesis conditions and predicting the effects of different variables on material properties, allowing for greater control in the synthesis of new materials, which is a way to save energy.

4.1.4 Computational methods and environmental sustainability

Consequently, computational techniques allow for the design and optimization of CNMs that are critical for eventual development. Furthermore, comprehensive evaluation of the synthesized products' influence on the environment is essential. Accordingly, researchers can identify the most eco-friendly and least emitting products to prioritize [15]. The combination of life cycle assessment (LCA) software with computational models has allowed the LCA of CNM manufacturing to be grounded in a holistic view of energy or materials consumption, emissions, and produced waste while following the principles of green chemistry and sustainable manufacturing [16]. Computation-driven studies permit the exploration of the new field of biomass-based CNMs as alternatives to traditional petrochemical-derived precursors. The feasibility and environmental viability of transformational processes based on these bio-derived materials can catalyze the creation of much more environmentally friendly technological approaches to the green synthesis of CNMs [17, 18]. Computational resources have transformed the field of sustainable CNM synthesis by promoting enhanced efficiency and sustainability due to property prediction, which leads to the identification of alternative pathways with the aid of DFT, MD simulation, or ML platforms.

4.1.5 Recent advancements in computational tools for materials synthesis

Materials science, particularly the synthesis of carbon-based nanomaterials, has been revolutionized by advanced computational approaches. Computational methods offer a way to speed up the design and optimization of new materials with desired properties, with higher predictive accuracy than traditional experimentation. For the preparation of CNMs, modeling approaches such as MD simulation, DFT, and ML are essential tools [19–21]. The goal of these methods is to engineer materials with better properties, design processes with optimized parameters, and reduce resource utilization to achieve sustainable nanomaterial synthesis.

4.2 Density functional theory and its applications

DFT is the cornerstone of modern computational materials science and quantum chemistry [22]. It provides a way to investigate the electronic structures of atomic, molecular, solid, and quantum nanostructures [23]. DFT is based on the seminal works of Hohenberg and Kohn from 1964 and was later expanded to the many-electron case by the Kohn–Sham approximation in 1965. To put it briefly, the central assumption of DFT is that the ground-state energy of a many-electron system can be represented as a function of the electron density and not a functional of the many-body wave function [24]. In doing so, DFT eliminates the need to consider the wave function, which is almost impossible to calculate in practice.

Density is a three-dimensional quantity (x, y, z); hence, no matter how many particles there are, DFT cuts through the problems in a scale-free way with no loss of accuracy.

DFT calculations are used to obtain information about the electronic structure, reactivity, and stability of carbon-based nanomaterials (CNMs). This information is used to design new materials in high-technology areas such as energy storage, electronics, semiconductors, catalysts, drugs, and biotechnology [25, 26]. The key information derived from DFT calculations includes band structure, density of states (DOS), projected density of states (PDOS), charge density distribution, adsorption energy calculation, and phonon dispersion [27]. These features influence the preparation and improvement of CNMs.

4.2.1 Applying the principles of density functional theory to carbon nanomaterials

The organizing principle of DFT is electron density, a far more accurate representation of electron behavior in matter than anything known in quantum mechanics. DFT solves the Schrödinger equation for a many-body system and provides fundamentally important information about a material's properties, including its band structure, DOS, adsorption energy, charge distributions, and electronic properties [28–30].

The Kohn–Sham equations are the basis of the practical use of DFT.

Kohn–Sham equation:

$$\left[-\frac{h^2}{2m} \nabla^2 + V_{\text{eff}}(r) \right] \psi_i(r) = \varepsilon_i \psi_i(r), \tag{4.1}$$

where $\psi_i(r)$ is the Kohn–Sham orbital for the ith electron and E_i is its corresponding eigenvalue.

$V_{\text{eff}}(r)$ is the effective potential, consisting of:
- an external potential $V_{\text{ext}}(r)$;
- the Hartree potential $V_H(r)V$ for classical electron–electron repulsion; and
- the exchange-correlation potential $V_{\text{xc}}(r)$, which incorporates all quantum many-body effects:

$$V_{\text{eff}}(r) = V_{\text{ext}}(r) + V_H(r) + V_{\text{xc}}(r). \tag{4.2}$$

The total electron density is then given by:

$$\rho(r) = \sum_i |\psi_i(r)|^2. \tag{4.3}$$

The challenge in DFT lies in accurately modeling the exchange-correlation energy, $E_{\text{xc}}[\rho]$, which is typically approximated using:
- the local density approximation (LDA);
- the generalized gradient approximation (GGA) (e.g. Perdew–Burke–Ernzerhof (PBE)); or

- hybrid functionals (e.g. B3LYP, HSE06), which incorporate a portion of exact exchange from Hartree–Fock theory.

DFT calculations are crucial for the study of various CNMs, including graphene, CNTs, and carbon quantum dots (CQDs).
- The electronic properties and stability of graphene can be controlled through functionalization with chemical groups.
- Indeed, predicting the chirality of CNTs can increase their electrical conductivity.
- The unique optical properties of CQDs can be customized for biological imaging and sensors, given that they can anticipate the quantum confinement effect.

4.2.2 Applications in carbon nanomaterials

It is worth considering that DFT is not an exact recipe, since one needs to predict some material parameters before experimental synthesis. A stable configuration is established, and reactive sites along with local energies for specific carbon formations are identified. Therefore, researchers can already plan their possible synthesis routes and reduce the costs of undesired experimental procedures. The most valuable target is green synthesis, in which researchers avoid large amounts of hazardous solvents and energy use. Thus, DFT can provide information about high-performance solvents or precursor materials used for CNT synthesis without depending on trial and error.

4.2.3 Electronic structure analysis

DFT is a very powerful simulation tool for electronic properties in terms of band structure and DOS and is used to estimate the conductance and semiconducting or metallic behavior of materials [31].

For instance, a distinct change in bandgap is observed between the armchair and zigzag edge configurations of graphene nanoribbons (GNRs) that leads to intramolecular magnetism and semiconducting behavior depending on the width of the GNR [32]. DFT calculates the bandgap and allows or forbids the band for the CNTs, which is highly significant for developing sensors, field emitters, and transistors. DFT shows that the metallic or semiconducting behavior of the CNT depends on its chirality and diameter, as shown in figure 4.2.

4.2.4 Band structure

The band structure indicates the allowed and unallowed bands of electrons, which is essential to predict electrical conductivity and semiconductor behavior.

The electrical conductivity of graphene is remarkable because it has a distinct Dirac cone at the K point of the Brillouin zone, as shown in figure 4.1 [33].

DFT calculations help researchers better understand how, for instance, doping or functionalization could induce a bandgap, turning a material from a semimetal into a semiconductor or insulator.

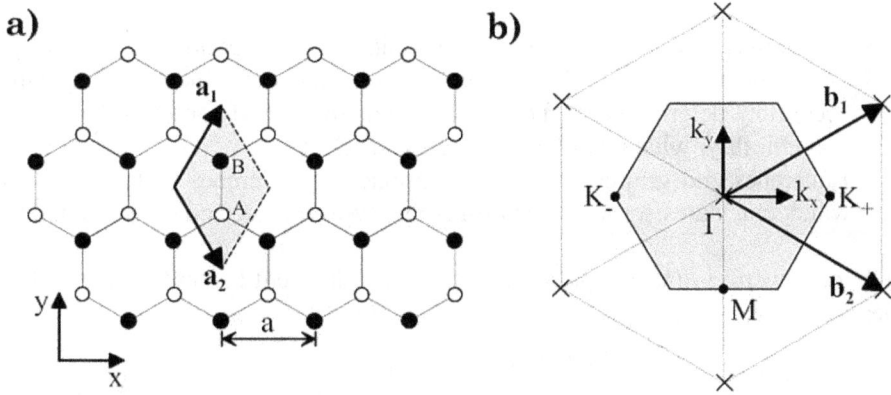

Figure 4.1. Graphene structure.

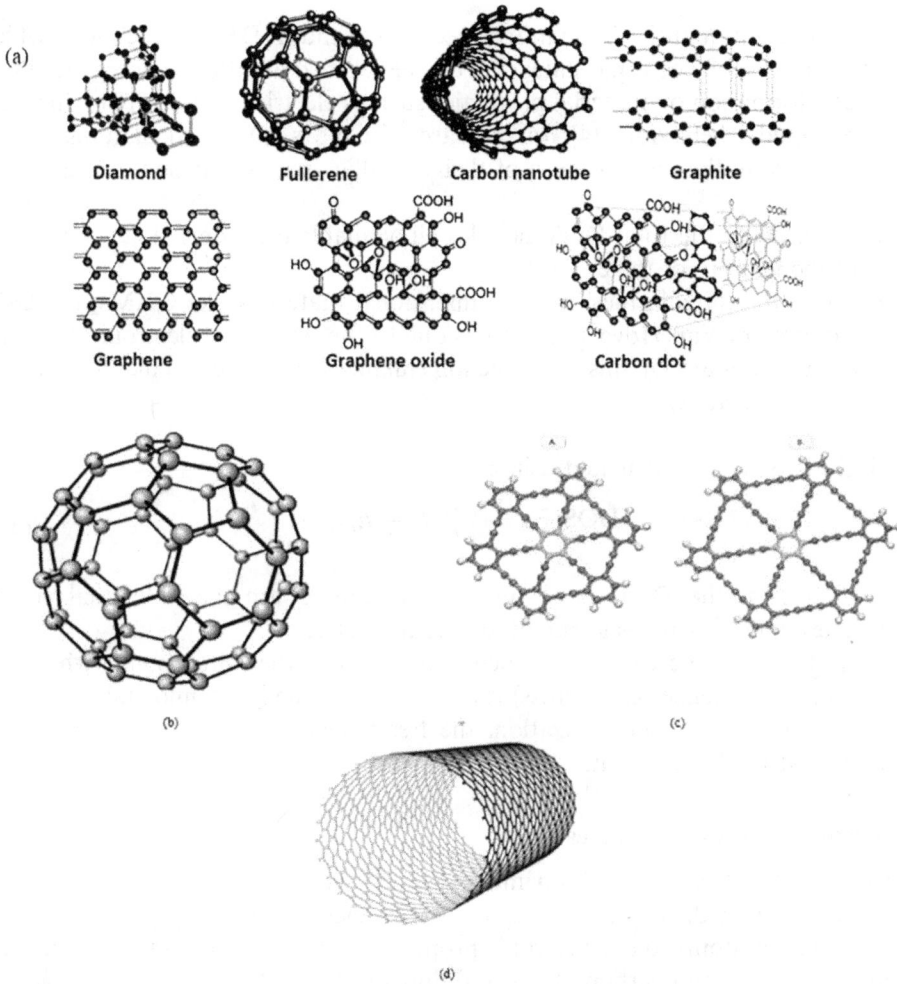

Figure 4.2. (a) Carbon nanostructures, (b) fullerene (C_{60}), (c) graphdiyne, and (d) a CNT.

- CNTs may vary in chirality and metallicity or semiconducting nature [14]. The band structure of CNTs that are metallic or semiconducting can be accurately determined by DFT. The bandgap depends on the tube diameter and chirality, which can be predicted with DFT.
- Fullerenes and graphdiyne: DFT is applied to fullerenes and other CNMs, which elucidates sensor mechanisms by revealing electronic behavior.

The band structure $E(k)$ is typically computed for different k-points in the Brillouin zone:

$$E(k) = \varepsilon(k) - E_{Fermi}, \tag{4.4}$$

where $\varepsilon(k)$ is the energy of the wave vector and E_{Fermi} is the Fermi level of the system.

4.2.5 Density of states

A very important quantity in quantum mechanics (and in DFT) is the DOS, which gives the number of electronic states that can be reached at different energy levels. It provides information on material properties such as electrical conductivity, magnetism, and optical behavior [34]. In the case of CNMs, DOS calculations play a significant role in the identification of their metallic, semiconducting, or insulating nature.

- Total DOS: the total DOS includes all available electronic states, regardless of the identity of the electron.
- Van Hove singularities: in low-dimensional materials like CNMs, the DOS can present Van Hove singularities due to the particular electronic structure of the material [35]; this phenomenon enhances properties such as absorbance and conductivity.

The DOS is typically calculated using:

$$\text{DOS}(E) = \sum_i \delta(E - E_i), \tag{4.5}$$

where δ delta is the Dirac delta function, E and E_i are the individual energy eigenvalues, and $\sum$ denotes summation over all states.

For instance, in the case of graphene, the DOS at the Dirac point (where the conduction and valence bands cross) is zero, so graphene is a semimetal. However, following doping or functionalization, the Fermi level may move and produce a finite DOS at the Dirac point.

4.2.6 Projected density of states

Analysis of the PDOS provides a more detailed view of the DOS, including the contributions from distinct atomic orbitals or species. Understanding the contribution of different atoms to the electronic properties of the material is important, as is distinguishing between carbon atoms and dopant atoms.

This assists scientists when studying the following:

- the states, e.g. *pzp_z*, ss, *pxp_x*, responsible for the development of conduction or valence bands;
- the effect of dopants, for example, nitrogen in graphene, or functional groups on the electronic structure and reactivity of materials.

DFT is used to calculate the PDOS of C and N atoms in N-doped graphene, which contributes to an increase in electrocatalytic activity by creating new states near the Fermi level. This visualization helps in the material design for specific applications by showing how nitrogen affects electronic states.

The PDOS is often calculated using:

$$\text{PDOS}_{\alpha}(E) = \sum_{i} \delta(E - E_i) <\psi_i|\alpha>^2, \qquad (4.6)$$

where α refers to the orbital type or atom of interest, and $\langle\psi_i|\alpha\rangle^2$ represents the overlap of the eigenfunction ψ_i with the orbital α.

4.2.7 Charge density and charge transfer

Charge density, which is the spatial distribution of electrons in a material, plays a crucial role in the material's electronic and chemical nature. First-principles studies using DFT provide a full mapping of this distribution. This provides vital information about the charge distribution around atoms or nearby functional groups, as well as how charge spreads or leaks across different parts of the material, such as into CNT edges or defect sites in graphene [36].

The DFT results provide a theoretical description of electron density changes across different regions of CNMs, offering insights into their electrical conduction and reactivity. In graphene, electron density is typically distributed uniformly in two dimensions; however, localized charge rearrangements can significantly alter its electronic response. The electron-deficient nature of nitrogen atoms induces a redistribution of electron density within graphene sheets, creating active sites for chemical reactions and modulating the overall electronic structure [37].

4.2.8 Charge redistribution

DFT calculates the electronic charge redistribution for CNMs that alter their electrical conductivity, optical properties, and catalytic performance in the presence of N, B, or S dopants [38]. N-doped graphene shows enhanced electrical conductivity due to the higher charge carrier density originating from nitrogen atoms, especially near the Fermi level [39]. Also, boron-doped graphene exhibits p-type conductivity, in which the boron atoms act as electron acceptors that reduce the electron density, leading to the creation of holes for hole conduction [40, 41].

4.2.9 Charge transfer mechanisms in catalysis

Charge transfer is an elementary step in catalysis and an important topic, since it also governs the interaction of a catalyst with reactants, particularly in electrocatalysis [42, 43]. DFT is used to investigate the charge transfer process in catalysis, enabling the development of more effective catalysts [44]. For energy storage and conversion devices (batteries and fuel cells), the use of DFT to predict electron transfer between catalysts and reactants has a significant impact. DFT calculations have been used to estimate the adsorption energy of CO_2 on the catalyst surface and the electronic properties of the catalyst in the CO_2 reduction reaction.

4.2.10 Phonon dispersion and vibrational properties

Lattice vibrations are described by phonons and are indispensable in understanding the thermal properties of materials because they define the oscillatory motion of atoms about their equilibrium positions [45–47]. Phonon dispersion plays a vital role in engineering thermal management plans in electronic devices and energy storage systems because it reflects the phenomenon that materials with low phonon scattering feature high thermal conductivity.

4.2.11 Thermal conductivity in graphene

The widely researched graphene possesses high thermal conductivity due to its strong carbon–carbon bonds and low mass, which play a crucial role in effective phonon propagation and the achievement of a high phonon group velocity. The phonon dispersion of graphene has a linear relationship in the low-frequency regime, indicating the effective transmission of the acoustic phonon [48]. In addition, defects and functionalization can reduce thermal conductivity, as suggested by DFT calculations.

4.2.12 Nanotubes and graphene sheets

Phonon dispersion has been used to study [49, 50] the effect of defects or functional groups on the thermal properties of systems such as CNTs, graphene sheets, etc. Chemical functionalization is one of the approaches most frequently employed to influence the properties of CNMs for their intended use, e.g. elevated solubility, introduced reactivity, and enhanced biocompatibility [10, 51]. The use of DFT for *in silico* calculations offers researchers insights into how various functional groups or heteroatoms (e.g. B, N, and S) influence electric properties (bandgap, charge density), magnetic features (especially in edge-functionalized graphene), and chemical activity (such as enhanced hydrogen adsorption for energy storage).

For example, carboxylated CNTs may greatly improve hydrophilicity, charge transfer capacity, and binding activity of CNTs with biomolecules or metal NPs. Also, DFT calculations for charge redistribution and binding energies have proven that N-graphene exhibits better catalytic activity than pristine graphene for ORR.

4.2.13 Advantages of density functional theory in sustainable carbon nanomaterial synthesis

- Resource efficiency: DFT allows the virtual screening of materials without the need to invest in expensive lab tests, reducing experimental overhead.

- Predictive design: this enables scientists to predict performance before synthesis, perfect for green and safe synthesis.
- Tailored properties: DFT directs atom-level design (e.g. defects, dopants), which is used to develop CNMs with the desired catalytic, electrical, or mechanical properties.
- Utilization of AI: properties calculated using DFT represent useful property training data sets for ML in the discovery of materials.

4.2.14 Software for density functional theory simulations

Several DFT-based simulation packages are widely used in CNM research, each offering unique advantages in terms of accuracy, computational cost, and user interface. Quantum ESPRESSO is a plane-wave-based, open-source tool; Vienna Ab-initio Simulation Package (VASP) is commercial software known for its high accuracy; Gaussian is widely used in chemistry, particularly for molecular CNMs; Material Studio is an open-source molecular modeling platform; and Quantum ATK is designed for atomistic simulations. Other notable tools include Visualisation for Electronic Structural Analysis (VESTA), Spanish Initiative for Electronic Simulations with Thousands of Atoms (SIESTA), ORCA, ABINIT, CASTEP, and BURAI-GUI. These packages are highly adaptable for various CNM applications such as structural optimization, phonon calculations, and molecular orbital analysis.

4.3 Molecular dynamics simulations and applications

During synthesis, the electronic properties of CNMs can be investigated using DFT, while their atomic and molecular-level behaviors can be explored through MD. MD simulations are performed by solving the equations of motion of a collection of atoms, enabling researchers to keep track of the positions, velocities, and accelerations of the atoms as a function of time [52, 53]. This approach describes how CNMs are generated, grow, and behave under different settings. In particular, by simulating the growth of CNMs, MD simulations are valuable for investigating the growth processes of carbon-based nanomaterials, which are important for describing the material properties during the growth. For instance, MD simulations can simulate the real-time growth of CNTs or graphene sheets. Through emulating these processes, researchers can tailor growth conditions, including temperature, precursors, and growth duration, which are known to affect the final properties of the material.

- Growth of CNTs: for CNTs, MD simulations may also shed light on the mechanism of CNT nucleation and elongation growth parameters on the chirality and electrical properties of CNTs [54, 55]. It is notable that a better understanding of how to control the CNT growth dynamics via MD might eventually allow for better control over the properties and structure of CNTs and could, therefore, be employed to improve their performance in devices, from field-effect transistors to supercapacitor applications.

- Folding of a graphene sheet: MD simulations could also help understand graphene sheet folding and wrinkling during synthesis [55, 56]. These imperfections or curvatures may account for the increased surface area and reactivity of graphene, which is important for its applications in catalysis and energy storage [24, 57, 58].

4.3.1 Optimizing synthesis conditions

Similarly, MD simulations can also be used to predict how CNMs will behave in response to a collection of external conditions (e.g. temperature, pressure, etc.) during synthesis. By mimicking these conditions, scientists should be able to determine the most efficient and cost-effective synthesis methods. For instance, MD simulations of CVD processes for the growth of CNTs could be used to determine the optimal CVD temperature [59] and precursor flows required to obtain the maximum CNT yield with the lowest number of defects.

Furthermore, MD simulations can be used to understand how precursor molecules interact with substrates at the atomic scale. This allows researchers to find the best catalyst or solvent for CNM growth, decreasing the need to employ expensive or toxic chemicals during synthesis.

Key equations:

The fundamental equation governing MD simulations is Newton's second law:

$$M_i \frac{d^2 r_i}{dt^2} = F_i, \tag{4.7}$$

where F_i denotes the force on an atom i, derived from the interatomic potential.

4.4 Machine learning in materials discovery

Recently, ML has been steadily gaining popularity in materials discovery, focusing on the synthesis of CNMs [60]. ML looks at large amounts of data to find patterns and correlations, making it critical in the analysis of complex, multifaceted data sets. ML can predict the material properties of CNMs, such as their electrical, mechanical, and thermal properties [61]. It can learn from both experimental and simulation data, accelerating the process of synthesis by correlating synthesis conditions.

- ML: these methods can expose hidden relationships between synthesis parameters and material properties to enable data-driven discovery [62, 63].
- Catalyst screening: high-throughput screening is an attractive application of ML in CNM synthesis, allowing the quick evaluation of candidate materials and the production of a shortlist for experimental validation [64–66].

In addition, ML enhances input resource use by predicting optimal synthesis conditions, minimizing material and energy consumption, and suggesting more efficient reactions or lower-temperature materials and reaction methods, with a concomitant reduction in the environmental impact of CNM production.

4.4.1 Applications of machine learning in the synthesis of carbon nanomaterials

ML models are used to predict the optimum conditions for the synthesis of GO and rGO to enhance their properties for supercapacitors and battery electrodes [67–69]. In addition, optical properties of CQDs can be predicted with machine-learned models and used as the input to a simple optical model of a CQD-sensitized solar cell, indicating that careful design is necessary to optimize CQD-based photovoltaic devices and biosensors.

4.5 Design of experiments in the synthesis of carbon nanomaterials

The design of experiments (DOE), a statistical approach, is used in materials science to efficiently structure and carry out experiments [70]. It is particularly helpful in CNM synthesis, since it facilitates the determination of optimal experimental conditions and effective parameters to attain the desired properties [7]. DOE provides an organized procedure for making decisions, which allows researchers to examine numerous variables that influence material properties, such as temperature, pressure, metal precursors, catalyst types, and reaction times. In DOE, the input variables are systematically varied in a controlled fashion so that the optimal process conditions leading to the desired final product performance are ascertainable for the stages of operation.

- Defining objectives: the objectives could be yield maximization, energy consumption minimization, or material properties optimization.
- Selecting factors and levels: choosing which subject to analyze and which considerations to include.
- Experiment design: choose a design type that corresponds to the aims of the experiment. General examples of design types used in DOE include full factorial design, fractional design, and central composite design (CCD).
- Interactive: implement the design, run some experiments, and collect data to see patterns, interactions, and the best combination/level of factors.
- Analysis of results: understand the connection between the input and output properties.

4.5.1 Response surface methodology

Statistical models such as response surface methodology (RSM) are also frequently employed to design experiments that maximize the synthesis of CNMs [71, 72]. RSM consists of fitting experimental values to a model equation that describes the relationship between the synthesis parameters and the properties of the material. The model, typically a second-degree polynomial model, includes linear and quadratic interactions between the factors [70]. This learned model is then used to predict the properties of materials grown under different synthesis conditions and guide researchers to the most suitable conditions to achieve the desired material.

4.5.1.1 Mathematical representation of response surface methodology

In an RSM, the links between Y (material properties) and the inputs X_1, X_2, ..., X_k (e.g., temperature, pressure, catalyst concentration) are described by a polynomial equation:

$$Y = \beta_0 + \sum_{i=1}^{k} \beta_i X_i + \sum_{i=1}^{k}\sum_{j=i}^{k} \beta_{ij} X_i X_j + \sum_{i=1}^{k} \beta_{ii} X_i^2 + \epsilon, \qquad (4.8)$$

where Y is the dependent variable (e.g. yield, purity, surface area); X_i are the independent variables (e.g. temperature, pressure, precursor concentration); β_0 is the intercept term; β_i, β_{ij}, and β_{ii} are the coefficients to be estimated; and ϵ is the error term. The terms involving X_i^2 and $X_i X_j$ represent the quadratic and interaction effects between the factors, respectively. Using RSM, it becomes possible to model and understand nonlinear effects and interactions between variables.

Steps used during RSM implementation [73]:
1. Factor selection: identify important factors of synthesis and their levels.
2. Design of experiments: use CCD or Box–Behnken design to study the factor space.
3. Model fitting: apply a second-order polynomial model to determine the effect of synthesis conditions on the material properties.
4. Optimization: use a calibrated model to predict optimal synthesis conditions.
5. Validation: test the model prediction experimentally.

Research into CNT production through chemical vapor deposition (CVD) has shown that RSM can successfully optimize the system to maximize yields and purity while limiting the consumption of toxic solvents and chemicals [74].

4.5.2 Multi-objective optimization in carbon nanomaterial synthesis

Multi-objective optimization is an advanced DOE approach that can address many conflicting objectives at the same time, such as maximizing material production while minimizing energy usage and evaluating the environmental impact of CNM synthesis [36]. In a multi-objective optimization problem, the aim is to optimize several objective functions $f_1, f_2, \ldots$ given the constraints of the system. The typical form of the multi-objective optimization problem is:

$$\text{Minimise/maximise } f_1(X), \quad f_2(X), \ldots, f_m(X), \qquad (4.9)$$

subject to:

$$g_j(X) \leqslant 0, \quad j = 1, 2, \ldots, \quad p$$

$$h_k(X) = 0, \quad k = 1, 2, \ldots, \quad q,$$

where $X = (X_1, X_2, \ldots, X_k)$ are the decision variables (e.g. synthesis parameters), f_1, $f_2, \ldots, f_m$ are the multiple objective functions to be optimized, $g_j(X)$ are inequality constraints (e.g. temperature limit, precursor availability), and $h_k(X)$ are equality constraints (e.g. fixed catalyst concentration).

Steps in multi-objective optimization [75]:
1. Define goals: increase purity, shorten reaction time, reduce solvent use, and optimize material performance.
2. Develop the optimization model: the model uses mathematical tools that include linear programming and genetic algorithms to rank all objectives and establish the optimal design parameters for the synthesis.
3. Apply optimization: the optimization algorithm is a way to search numerous solutions (i.e. different synthesis parameters).
4. Use Pareto GUI analysis: it is easier to extract the best synthesis conditions from a visual representation of the trade-off between competing objectives.

A study of graphene oxide reduction involved a multi-objective optimization approach aimed at minimizing the consumption of the reducing agent, enhancing material performance, and reducing both waste generation and energy usage [76]. A sample of CNTs prepared by CVD was studied using RSM to improve the synthesis of CNTs, focusing on the effects of temperature, precursor concentration, and catalyst type [76, 77]. The research determined the optimum growth conditions for yield and quality with minimal environmental impacts. Another study utilized the DOE approach to optimize liquid-phase graphene production by adjusting the concentrations of the agents, the type of solvent, and the length of sonication [78]. The RSM model improved the scale-up of synthesis and reduced the consumption of harmful chemicals, which demonstrated the effectiveness of the DOE approach. In another study, bio-imaging-optimized CQDs were synthesized using temperature, precursor concentration, and reaction time as variables to systematically tune photoluminescence and biocompatibility, aided by RSM [79].

4.5.3 Advantages of design of experiments

DOE approaches optimize synthesis parameters to minimize raw material use, enhance energy efficiency, and reduce the generation of harmful byproducts and toxic solvents, thereby promoting environmentally friendly CNM production. By providing explicit guidelines for optimal synthesis conditions, DOE shortens experimental duration and time to market. Additionally, the identified methods exhibit improved scalability, making them suitable for industrial-scale CNM production.

4.5.3.1 Optimizing time, energy, and resources in the synthesis of carbon nanomaterials

In most cases, the classical CNM synthesis processes are the results of costly, labor-intensive, and ineffective trial-and-error methods. Computational approaches have revolutionized process design, offering the advantages of time and energy savings, material waste reduction, and, as a consequence, improved overall efficiency [80].

Using computational tools, researchers can predict and optimize synthesis parameters without conducting physical experiments, which speeds up the process and promotes daily work while focusing on sustainability [81]. Computational tools save time, labor, and reagents by streamlining reaction conditions, avoiding deleterious byproducts, and increasing material yields [82].

4.5.3.2 Minimizing duration by pre-synthesis predictions

In the course of synthesizing CNMs, computational techniques reduce the time required for experimental tests. Therefore, investigators can focus their attention on finding the optimal conditions for synthesis. This cyclic process may take weeks, months, or even longer, depending on the complexity of the content and the number of factors. Computational simulations, including DFT and MD, enable the prediction of material properties and reaction conditions in silico before experimental studies are undertaken [83]. Such simulations aid scientists in determining feasible synthesis routes, reducing the number of experimental attempts, and shortening the development timeline for CNMs, thus enabling fast iteration [84, 85]. It has been shown by several researchers that the use of computational tools, such as DFT, can reduce the synthesis temperature necessary to produce good-quality graphene and therefore decrease the trial-stage experiments in the development of scalable graphene manufacturing processes compared to conventional high-temperature methods. Computational approaches predict the optimal reaction conditions for CNM synthesis, helping to avoid unnecessary experiments and achieve higher efficiency for faster results.

4.5.3.3 Improving energy efficiency in synthesis processes

Computational methods increase energy savings in the production of CNMs in comparison with traditional methods, which are energy-consuming and require high temperatures and extended reaction times. This results in higher costs and a significant environmental footprint, particularly in terms of carbon emissions [86]. Computational methods, such as DFT-based calculations, can predict energy-efficient pathways to form CNMs, which reduce energy expenses and provide better-quality materials. DFT has also shown that low-temperature methods for graphene formation are possible in the presence of appropriate catalysts and precursors [87]. By simulating the reaction conditions, optimal parameters can be obtained from the DFT results, avoiding the low efficiency of experiments or the high energy consumption caused by the use of high temperatures.

Computational simulations enable the discovery of low-energy synthesis strategies (free pathways or agents), such as hydrothermal synthesis, which could considerably reduce energy costs in advance, avoiding time-consuming experimental processes for discovering new pathways. Computational simulations have predicted energy-efficient methods to control CNT orientation and diameter during manufacture, reducing power consumption and improving manufacturing procedures [88]. The computational methods reported can essentially enable a significant reduction in energy utilization in CNM synthesis by predicting the best reaction conditions, material generation, and structural integration, thereby leading to an overall reduction in energy footprint.

4.5.3.4 Reducing material waste and byproducts

Conventional synthesis approaches, such as those used for CNM synthesis, commonly result in substantial waste of precursor materials, solvents, catalysts, and byproducts, which aggravates environmental pollution [76, 89]. Computational approaches, including RSM and computational modeling, help control the amount of material used in studies of CNM synthesis, thereby reducing waste. Such approaches are especially useful for precious and rare resources, for which a decrease in consumption is imperative for the economy to remain viable and sustainable. Byproduct generation can be predicted by DFT and MD simulations, and alternative pathways, such as green chemistry mechanisms with nontoxic or biodegradable solvents and reagents, can also be explored to study sustainable processes and minimize undesirable byproducts [90, 91]. Computational methods aid in optimizing the synthesis of CQDs in terms of solvent and reaction conditions, reducing the number of harmful byproducts (including the anticipated catalyst-to-precursor ratio) and the overall expense of the synthesis [4, 92]. Computational processes enable scientists to control material usage, predict byproduct formation, and guide sustainable options to minimize overall waste and environmental impact.

4.5.3.5 Computational instruments in resource-efficient synthesis

Green manufacturing leverages computational methods to optimize the synthesis of CNMs by analyzing precursor requirements and refining reaction conditions to maximize yield while minimizing resource consumption and environmental impact [93]. Computational methods can also find energy-efficient pathways for product synthesis by modeling reactions, e.g. low-temperature CVD processes that could replace heat for graphene production [94]. From a computational point of view, reactions can also be optimized by exploring the best conditions (pressure, temperature, solvent) to enhance material yield and resource efficiency in CNM synthesis. Furthermore, computational methods enable the transformation of CNM synthesis by reducing energy input, cutting waste, and maximizing resources, thereby speeding up the process and improving its greenness.

Computational simulation and design facilitate the production of nanomaterials, help save time and resources, and promote sustainability and efficiency in green technology and industrial processing. Recent advances in computational methods enhance resource utilization efficiency, empowering researchers to design environmentally friendly CNMs that are essential for meeting the planet's future sustainability needs.

4.6 Applications of computationally synthesized carbon nanomaterials

Owing to their high surface area, electrical conductivity, mechanical strength, and biocompatibility, CNMs have been receiving significant attention. Computation has raised design and performance to new heights, expediting remarkable progress in the fields of energy, catalysis, biology, and environmental science.

4.6.1 Energy storage applications

CNTs (a type of CNM) are widely used in energy storage systems (batteries and supercapacitors) due to their high electrical conductivity, large surface area, and high mechanical strength [17, 31, 95, 96]. The theoretical approach, using DFT and/ or MD methods, is employed in predicting the behavior of these materials, as it is helpful in revealing their structural relationships and properties and improving their electrochemical performance [97]. The computed enhancement of graphene-based materials has even promoted the design of supercapacitors, CNTs, and their hybrids [55, 98]. Graphene oxide (GO) and reduced graphene oxide (rGO) have been presented as promising anode materials for electrochemical energy storage (Li-ion batteries), offering high capacity and cycling stability when integrated with computationally designed nanostructures [93]. Due to their excellent conductivity and high specific surface area, CNTs have also been extensively investigated for use as lithium-ion battery anode materials [92]. Computational models can be used to find stable positions of CNTs, which allow batteries to be more efficient and have longer lifetimes. MD simulations are used to investigate the mechanical properties of CNT supercapacitors under various conditions, including mechanical stress, mechanical loading, shielding, and thermal treatment. The energy storage performance of these new hybrid materials (such as graphene/CNT composites) is better than or equivalent to that of traditional energy devices [99].

4.6.1.1 Next-generation energy storage

Meanwhile, scientists are using ML to analyze experimental data to advance the design of solid-state batteries and supercapacitors and to predict as-yet-unknown carbon-based materials for energy storage.

Catalysis: CNMs such as CNTs and graphene are pivotal in catalytic reactions such as carbon dioxide reduction, hydrogenation, and oxygen reduction [16, 100]. Computational methods can also be used to optimize these catalysts, making them more efficient, cost-effective, and environmentally friendly.

CO_2 reduction and hydrogenation: DFT calculations have been utilized to study CO_2 reduction on CNTs. Doping with metal or nonmetal atoms can be used to enhance catalytic performance. Catalysts are designed to create products through the reduction of CO_2 into valuable compounds that can help capture and store carbon. The computational optimization of graphene-based materials for hydrogenation reactions enhances the performance of chemical synthesis by tuning the active sites on the graphene plane [101].

Electrocatalysis: graphene and CNTs have been computationally designed to obtain the best electron transport for oxygen reduction and hydrogen evolution, using DFT-based calculations to realize cheap catalysts for fuel cells and hydrogen production [47, 102]. Computational work attempts to improve the catalytic properties of CNTs by defect engineering or doping with heteroatoms, as well as by tuning the placement and concentration of these modifications for maximum effect [38, 103].

4.6.2 Biological applications

CNMs have shown potential in biological applications, including drug discovery, biosensors, and imaging, due to their biocompatible nature, large surface area, and capability to be modified by various biomolecules [104, 105].

Drug discovery: optimizing CNTs and GO for drug delivery via DFT and MD simulations provides a greater ability to increase drug loading and release, predict their toxicity, and aid in functionalization with peptides, antibodies, or targeting ligands, which contribute to diminishing side effects and improving drug efficiency [106].

Biosensing: this involves the computational design of carbon quantum dots and other graphene-based materials for sensing biomolecules such as DNA, proteins, and viruses. Their unique optical properties support fluorescent biosensing, which can be adapted to build sensitive biosensors for medical diagnostics, environmental control, and food safety [107, 108].

4.6.3 Applications in environmental science

Due to their immense surface area and functionalization, CNMs are considered promising materials for applications in the environmental sciences, including water purification, pollutant removal, and water sensing.

Water filtration and purification: computational modeling also enables the design of carbon-based nanomaterials for water filtration devices, such as GO and CNTs [36, 109–112]. Such materials are tuned for adsorption, and DFT simulations suggest their interactions and enhance their adsorption/filtration ability [113].

Environmental monitoring: the computational design approach for CNMs, including carbon quantum dots and CNTs, could be used for environmental monitoring. These agents can also be employed to detect pollutants, such as toxic metals or volatile organic compounds, which can be analyzed in real time using an online situational monitoring system [29, 114]. Computational approaches help in assessing the stability and repeated use of CNMs, thereby preventing the fast degradation of CNMs in environmental applications.

4.7 Challenges and future prospects

The development of CNMs has been greatly enhanced by computational approaches; however, several limitations need to be addressed, such as limitations in current methodologies, differences between theoretical predictions and experiments, and the requirement for green synthetic techniques. In this section, we discuss the limitations of the computational synthesis of CNMs; the roles of AI, ML, and quantum computing; and the circular economy model for the sustainable and efficient production of CNMs in the future.

4.7.1 Challenges in computational carbon nanomaterial synthesis

Despite recent advances in computational methods for designing CNM synthesis, challenges remain, particularly challenges related to computational scalability, the complex energy landscapes of many CNMs, and the limited ability of current models to accurately reflect experimental outcomes [115]. Computational techniques such as DFT, MD, and Monte Carlo simulations are widely used to design CNMs but are expensive, requiring significant amounts of resources [116]. CNMs are complex systems, so it is difficult to simulate them, and high-performance computers or supercomputers may not be available. Even with algorithms that have continued to evolve, knowledge of computational tools and materials science is necessary for the accurate construction and interpretation of CNM models [117]. The gulf between theoretical predictions and experimental observations in the realm of CNM synthesis constitutes a significant challenge, often owing to overly simplified simulations and unpredictable factors, as well as an inability to account for impurities, defects, and environmental influences [118]. Computational approaches predict the properties of CNM formation, but the experimental situation can be different. Filling this gap will require improved models and collaboration among theorists and experimentalists.

4.7.2 AI and machine learning integration: the future of carbon nanomaterial synthesis

AI and ML are integrated into computational modeling to advance the predictions of CNM synthesis and enable accurate and faster predictions of material properties, synthesis conditions, and performance [115]. AI and ML algorithms can rapidly handle extensive data sets, which allows for high-throughput screening of prospective material candidates. By training models on existing experimental data, we can predict the properties of new CNMs and reduce the time needed for their discovery [119]. ML-based models optimize CNM synthesis conditions based on previously conducted experiments that evaluate potential parameters, including but not limited to temperature, pressure, and precursor concentration, thus reducing inefficiencies and identifying successful pathways to producing high-quality materials [91]. ML enables predictive modeling and multi-objective optimization in the synthesis of CNMs, allowing researchers to balance conflicting objectives such as maximizing material yield, minimizing energy consumption, and reducing byproduct formation. AI and ML also facilitate the discovery of new CNMs by identifying hidden patterns within large data sets. Compared to conventional material design approaches, this integration significantly enhances synthesis efficiency and promotes more sustainable manufacturing processes [119].

4.7.3 Quantum computing: unleashing new possibilities

Quantum computing has been employed to simulate complex materials, such as carbon-based nanomaterials (CNMs), at a level of detail that surpasses the capabilities of classical computing. By enabling the simultaneous processing of vast amounts of data, quantum computing allows for highly accurate and comprehensive simulations. This, in turn, facilitates the design of CNMs with tailored properties and the optimization of their structures for specific applications, particularly in enhancing their electronic performance [9]. Quantum computers can simulate complex systems, including large CNMs composed of thousands of atoms. This capability offers new insights into atomic-level interactions and opens up possibilities for discovering novel CNMs with diverse properties. By enabling the parallel simulation of numerous candidate materials, quantum computing has the potential to revolutionize material discovery, significantly reducing design time and empowering researchers to efficiently identify and explore promising options [121]. Quantum computing is expected to be more efficient than conventional computing methods and will provide new design and synthesis methods for CNMs.

4.7.4 Circular economy approaches: sustainability across the carobn nanomaterial life cycle

Circular economy principles, including recycling and reuse, along with minimal environmental impact during the CNM life cycle, are key to sustainability [122]. A circular economy would maximize the recycling and reuse of materials, such as CNTs and graphene, from electronic waste and spent batteries. Computational methods can be employed to find efficient reprocessing techniques. The concept of sustainability benefits both the environment and CNM production by introducing circular economy principles. These include designing materials with life cycle considerations, using biodegradable and non-toxic components, optimizing synthesis processes, and minimizing resource inputs and waste outputs, ultimately making production more sustainable [123]. LCA scrutiny of the environmental impact of a material during its service life has enabled computational work to predict the environmental impact of CNMs, promoting the utilization of eco-friendly materials and processes [124]. The design and synthesis of CNMs following circular economy principles can significantly reduce their environmental harm and contribute to sustainable development.

4.8 Conclusion

Computational tools, e.g. DFT, MD simulations, and ML, have revolutionized the synthesis of CNMs, enabling materials to be designed with tailored properties for different applications and supporting the development of more sustainable synthesis methods.

4.8.1 Tailored properties and customization of carbon nanomaterials

The use of computational methods in CNM synthesis allows materials to be designed *in silico*, enabling scientists to explore design parameters before engaging in lab synthesis. The aforementioned design process yields improved properties, such as enhanced electrical conductivity, mechanical strength, and chemical stability, which are essential for applications in energy storage, catalysis, and environmental remediation. DFT and MD are two techniques commonly used for the study of CNTs (CNMs) and graphene. DFT enables the study of electronic structure at the atomic level, while MD simulations allow scientists to probe physical properties and also tune the synthesis parameters in order to achieve specific properties, making these computational methods essential for many applications. Material discovery is undergoing a revolution due to ML, and the search for materials that fit a benchmark scenario has been transformed through the exploration of vast collections of materials data, the recognition of patterns, and fast screening of candidate materials to be applied for specific purposes, with enormous savings in time and resources.

4.8.2 Optimizing sustainability in synthesis

The importance of materials science in sustainability lies in its power to reduce energy use, decrease waste, and lessen the ecological footprint. Computational tools optimize the production routes of CNMs by increasing energy efficiency and reducing environmental impact. *In silico* synthesis simulations can minimize the use of resources and energy by searching for energy-efficient ways to grow CNTs or synthesize graphene, leading to synthesis with a smaller environmental footprint. Computer-aided tools such as DOE, RSM, or multi-objective optimization are valuable for balancing yield, material quality, and the sustainability of environmentally friendly CNM synthesis paths, considering energy consumption, cost, and the associated environmental consequences.

4.8.3 Advances in computational methods: a paradigm shift

Computational science, particularly the application of AI and ML, has opened new avenues for the synthesis and design of CNMs, enabling more accurate predictions in materials research. Advances in quantum computing further enhance the simulation of materials at the atomic and molecular scales, allowing for deeper, real-world insights into the properties and behavior of complex CNMs. The integration of quantum computing and AI into materials science offers a powerful framework for optimizing CNM synthesis, facilitating the design of materials with exceptional properties while minimizing environmental and economic costs.

4.8.4 Circular economy and sustainable production of carbon nanomaterials

Computational models are integrating circular economy principles into the synthesis of CNMs. The objectives of this strategy are waste reduction, recycled material recovery, and reduced environmental impact in product manufacturing. In search of sustainable materials that may be efficiently transformed into high-quality CNMs, increasing research attention has been paid to the development of new CNM synthesis methods based on sustainable carbon sources, such as biomass. To contribute to a circular economy, simulations may also construct CNMs that are easy to reclaim and reuse. For a more sustainable future and to reduce waste, one could, in the future, imagine computational tools that can track the full life cycle of CNMs and thus maximize resource efficiency.

References and further reading

[1] Ayanda O S, Mmuoegbulam A O, Okezie O, Durumin Iya N I, Mohammed S A E, James P H, Muhammad A B, Unimke A A, Alim S A and Yahaya S M 2024 Recent progress in carbon-based nanomaterials: critical review *J. Nanopart. Res.* **26** 106

[2] Díez-Pascual A M 2021 *Carbon-Based Nanomaterials* (MDPI) p 7726

[3] Rao N, Singh R and Bashambu L 2021 Carbon-based nanomaterials: synthesis and prospective applications *Mater. Today Proc.* **44** 608–14

[4] Konstantopoulos G, Koumoulos E P and Charitidis C A 2022 Digital innovation enabled nanomaterial manufacturing; machine learning strategies and green perspectives *Nanomaterials* **12** 2646

[5] Hippalgaonkar K, Li Q, Wang X, Fisher III J W, Kirkpatrick J and Buonassisi T 2023 Knowledge-integrated machine learning for materials: lessons from gameplaying and robotics *Nat. Rev. Mater.* **8** 241–60

[6] Manolis G D, Dineva P S, Rangelov T and Sfyris D 2021 Mechanical models and numerical simulations in nanomechanics: a review across the scales *Eng. Anal. Boundary Elem.* **128** 149–70

[7] Fajimi L I, Oboirien B O and Yaqub Z T 2024 Modeling and optimization of nano-materials production processes *Nanomaterials for Sustainable Hydrogen Production and Storage* (Boca Raton, FL: CRC Press) 97–122

[8] Qasim M, Clarkson A N and Hinkley S F 2023 Green synthesis of carbon nanoparticles (CNPs) from biomass for biomedical applications *Int. J. Mol. Sci.* **24** 1023

[9] Chiminelli A, Radović I, Fasano M, Fantoni A, Laspalas M, Kalinić A, Provenzano M and Fernandes M 2024 Modeling carbon-based nanomaterials (CNMs) and derived composites and devices *Sensors* **24** 7665

[10] Badar M S, Shamsi S, Ahmed J and Alam M A 2022 Molecular dynamics simulations: concept, methods, and applications *Transdisciplinarity* (Berlin: Springer) 131–51

[11] Gajaria T K, Chodvadiya D and Jha P K 2021 Density functional theory investigation of thermal conductivity in α-CN and α-CP monolayers: implications for thermal management of electronic devices *ACS Appl. Nano Mater.* **4** 4474–83

[12] Yadav M D, Joshi H M, Sawant S V, Dasgupta K, Patwardhan A W and Joshi J B 2023 Advances in the application of carbon nanotubes as catalyst support for hydrogenation reactions *Chem. Eng. Sci.* **272** 118586

[13] Choudhary K, DeCost B, Chen C, Jain A, Tavazza F, Cohn R, Park C W, Choudhary A, Agrawal A and Billinge S J 2022 Recent advances and applications of deep learning methods in materials science *NPJ Comput. Mater.* **8** 59

[14] Tang D-M, Erohin S V, Kvashnin D G, Demin V A, Cretu O, Jiang S, Zhang L, Hou P-X, Chen G and Futaba D N 2021 Semiconductor nanochannels in metallic carbon nanotubes by thermomechanical chirality alteration *Science* **374** 1616–20

[15] Aithal P S and Aithal S 2022 Opportunities and challenges for green and eco-friendly nanotechnology in twenty-first century *Sustainable Nanotechnology: Strategies, Products, and Applications* (Hoboken, NJ: Wiley) 31–50

[16] Martínez-Ramón N, Calvo-Rodríguez F, Iribarren D and Dufour J 2024 Frameworks for the application of machine learning in life cycle assessment for process modeling *Clean. Environ. Syst.* **14** 100221

[17] Verma C, Chauhan D S, Aslam R, Banerjee P, Aslam J, Quadri T W, Zehra S, Verma D K, Quraishi M A and Dubey S 2024 Principles and theories of green chemistry for corrosion science and engineering: design and application *Green Chem.* **26** 4270–357

[18] Lesch R, Visser E D, Seroka N S and Khotseng L Biomass-derived carbon-based nanomaterials: current research, trends, and challenges 1–43

[19] Sibanda D, Oyinbo S T and Jen T-C 2022 A review of atomic layer deposition modelling and simulation methodologies: density functional theory and molecular dynamics *Nanotechnol. Rev.* **11** 1332–63

[20] Liu W, Wu Y, Hong Y, Zhang Z, Yue Y and Zhang J 2022 Applications of machine learning in computational nanotechnology *Nanotechnology* **33** 162501

[21] Teale A M, Helgaker T, Savin A, Adamo C, Aradi B, Arbuznikov A V, Ayers P W, Baerends E J, Barone V and Calaminici P 2022 DFT exchange: sharing perspectives on the workhorse of quantum chemistry and materials science *Phys. Chem. Chem. Phys.* **24** 28700–81

[22] Makkar P and Ghosh N N 2021 A review on the use of DFT for the prediction of the properties of nanomaterials *RSC Adv.* **11** 27897–924

[23] Fedotov A, Vakhrushev A, Severyukhina O, Sidorenko A, Savva Y, Klenov N and Soloviev I 2021 Theoretical basis of quantum-mechanical modeling of functional nano-structures *Symmetry* **13** 883

[24] Luo X, Zheng H, Lai W, Yuan P, Li S, Li D and Chen Y 2023 Defect engineering of carbons for energy conversion and storage applications *Energy Environ. Mater.* **6** e12402

[25] Kothandam G, Singh G, Guan X, Lee J M, Ramadass K, Joseph S, Benzigar M, Karakoti A, Yi J and Kumar P 2023 Recent advances in carbon-based electrodes for energy storage and conversion *Adv. Sci.* **10** 2301045

[26] Liao X, Lu R, Xia L, Liu Q, Wang H, Zhao K, Wang Z and Zhao Y 2022 Density functional theory for electrocatalysis *Energy Environ. Mater.* **5** 157–85

[27] Singh P and Harbola M K 2021 Density-functional theory of material design: fundamentals and applications-I *Oxf. Open Mater. Sci.* **1** itab018

[28] Schmidt M 2022 Power functional theory for many-body dynamics *Rev. Mod. Phys.* **94** 015007

[29] Singh S and Kaur I 2020 Bandgap engineering in armchair graphene nanoribbon of zigzag-armchair-zigzag based nano-FET: a DFT investigation *Phys. E* **118** 113960

[30] Al-Mahayni H, Wang X, Harvey J P, Patience G S and Seifitokaldani A 2021 Experimental methods in chemical engineering: density functional theory *Can. J. Chem. Eng.* **99** 1885–911

[31] Tang D-M, Cretu O, Ishihara S, Zheng Y, Otsuka K, Xiang R, Maruyama S, Cheng H-M, Liu C and Golberg D 2024 Chirality engineering for carbon nanotube electronics *Nat. Rev. Electr. Eng.* **1** 149–62

[32] Shah K A and Parvaiz M S 2025 Recent advances in modelling of nanoscale electronic devices and their applications *Integr. Ferroelectr.* **241** 1–23

[33] Jana S, Bandyopadhyay A, Datta S, Bhattacharya D and Jana D 2021 Emerging properties of carbon-based 2D material beyond graphene *J. Phys. Condens. Matter* **34** 053001

[34] Toriyama M Y, Ganose A M, Dylla M, Anand S, Park J, Brod M K, Munro J M, Persson K A, Jain A and Snyder G J 2022 How to analyse a density of states *Mater. Today Electron.* **1** 100002

[35] Igoshev P and Irkhin V Y 2022 Giant density-of-states van Hove singularities in the face-centered cubic lattice *Phys. Lett.* A **438** 128107

[36] Vir Singh M, Kumar Tiwari A and Gupta R 2023 Catalytic chemical vapor deposition methodology for carbon nanotubes synthesis *ChemistrySelect* **8** e202204715

[37] Singh M B, Jain P, Mohammad F, Singh P, Bahadur I and Abedigamba O P 2024 Significant Increase in the dipole moment of graphene on functionalization: DFT calculations and molecular dynamics simulations *ACS Omega* **9** 16458–68

[38] Boshir Ahmed M, Alom J, Hasan M S, Asaduzzaman M, Rahman M S, Hossen R, Abu Hasan Johir M, Taufiq Alam M, Zhou J L and Zhu Y 2023 General doping chemistry of carbon materials *ChemNanoMat* **9** e202200482

[39] Bie C, Yu H, Cheng B, Ho W, Fan J and Yu J 2021 Design, fabrication, and mechanism of nitrogen-doped graphene-based photocatalyst *Adv. Mater.* **33** 2003521

[40] Ubhi M K, Kaur M, Grewal J K and Sharma V K 2023 Phosphorus-and boron-doped graphene-based nanomaterials for energy-related applications *Materials* **16** 1155

[41] Ngidi N P, Ollengo M A and Nyamori V O 2020 Tuning the properties of boron-doped reduced graphene oxide by altering the boron content *New J. Chem.* **44** 16864–76

[42] Sun S, Zhang Y, Shi X, Sun W, Felser C, Li W and Li G 2024 From charge to spin: an in-depth exploration of electron transfer in energy electrocatalysis *Adv. Mater.* **36** 2312524

[43] Yan Y, Nashath F Z, Chen S, Manickam S, Lim S S, Zhao H, Lester E, Wu T and Pang C H 2020 Synthesis of graphene: potential carbon precursors and approaches *Nanotechnol. Rev.* **9** 1284–314

[44] Srivastava G P 2022 *The Physics of Phonons* (Boca Raton, FL: CRC Press)

[45] Pallikara I, Kayastha P, Skelton J M and Whalley L D 2022 The physical significance of imaginary phonon modes in crystals *Electron. Struct.* **4** 033002

[46] Balandin A A 2020 Phononics of graphene and related materials *ACS Nano* **14** 5170–8

[47] Zhu J and Mu S 2020 Defect engineering in carbon-based electrocatalysts: insight into intrinsic carbon defects *Adv. Funct. Mater.* **30** 2001097

[48] Wei Z, Duan Z, Kan Y, Zhang Y and Chen Y 2020 Phonon energy dissipation in friction between graphene/graphene interface *J. Appl. Phys.* **127** 015105

[49] Xie Y and Wang X 2023 Thermal conductivity of carbon-based nanomaterials: deep understanding of the structural effects *Green Carbon* **1** 47–57

[50] Ramburrun P, Khan R A and Choonara Y E 2022 Design, preparation, and functionalization of nanobiomaterials for enhanced efficacy in current and future biomedical applications *Nanotechnol. Rev.* **11** 1802–26

[51] Burdanova M G, Kharlamova M V, Kramberger C and Nikitin M P 2021 Applications of pristine and functionalized carbon nanotubes, graphene, and graphene nanoribbons in biomedicine *Nanomaterials* **11** 3020

[52] Alavi S 2020 *Molecular Simulations: Fundamentals and Practice* (New York: Wiley.)

[53] Wang L, Tricard N, Chen Z and Deng S 2025 Progress in computational methods and mechanistic insights on the growth of carbon nanotubes *Nanoscale* **17** 11812–63

[54] Qiu L and Ding F 2021 Understanding single-walled carbon nanotube growth for chirality controllable synthesis *Acc. Mater. Res.* **2** 828–41

[55] Zhang C, Lu C, Pei L, Li J and Wang R 2021 The wrinkling and buckling of graphene induced by nanotwinned copper matrix: a molecular dynamics study *Nano Mater. Sci.* **3** 95–103

[56] Jain V and Kandasubramanian B 2020 Functionalized graphene materials for hydrogen storage *J. Mater. Sci.* **55** 1865–903

[57] Wang J, Shi H, Wang W, Xu Z, Hong C, Xue Y and Tian F 2022 Defect engineering of graphynes for energy storage and conversion *Chem. Eng. J.* **432** 133617

[58] Gakis G P, Termine S, Trompeta A-F A, Aviziotis I G and Charitidis C A 2022 Unraveling the mechanisms of carbon nanotube growth by chemical vapor deposition *Chem. Eng. J.* **445** 136807

[59] Cai J, Chu X, Xu K, Li H and Wei J 2020 Machine learning-driven new material discovery *Nanoscale Adv.* **2** 3115–30

[60] Li Y *et al* 2025 Transforming the synthesis of carbon nanotubes with machine learning models and automation *Matter* **8** 101913

[61] Vivanco-Benavides L E, Martínez-González C L, Mercado-Zúñiga C and Torres-Torres C 2022 Machine learning and materials informatics approaches in the analysis of physical properties of carbon nanotubes: a review *Comput. Mater. Sci.* **201** 110939

[62] Ryu B, Wang L, Pu H, Chan M K and Chen J 2022 Understanding, discovery, and synthesis of 2D materials enabled by machine learning *Chem. Soc. Rev.* **51** 1899–925

[63] Datar A, Lyu Q and Lin L C 2023 Machine learning-aided discovery of nanoporous materials for energy-and environmental-related applications *AI-Guided Design and Property Prediction for Zeolites and Nanoporous Materials* (Hoboken, NJ: Wiley) 283–318

[64] Lombardo T, Duquesnoy M, El-Bouysidy H, Årén F, Gallo-Bueno A, Jørgensen P B, Bhowmik A, Demortière A, Ayerbe E and Alcaide F 2021 Artificial intelligence applied to battery research: hype or reality? *Chem. Rev.* **122** 10899–969

[65] Paul J T 2020 *Computational Discovery and Characterization of Low-Dimensional Materials* (Gainesville, FL: University of Florida)

[66] Saad A G, Emad-Eldeen A, Tawfik W Z and El-Deen A G 2022 Data-driven machine learning approach for predicting the capacitance of graphene-based supercapacitor electrodes *J. Energy Storage* **55** 105411

[67] Yogesh G K, Nandi D, Yeetsorn R, Wanchan W, Devi C, Singh R P, Vasistha A, Kumar M, Koinkar P and Yadav K 2025 A machine learning approach for estimating supercapacitor performance of graphene oxide nano-ring based electrode materials *Energy Adv.* **4** 119–39

[68] Liu X, Ji D, Jin X, Quintano V and Joshi R 2023 Machine learning assisted chemical characterization to investigate the temperature-dependent supercapacitance using Co-rGO electrodes *Carbon* **214** 118342

[69] Antony J 2023 *Design of Experiments for Engineers and Scientists* (Amsterdam: Elsevier)

[70] Lamberti F, Mazzariol C, Spolaore F, Ceccato R, Salmaso L and Gross S 2022 Design of experiment: a rational and still unexplored approach to inorganic materials' synthesis *Sustain. Chem.* **3** 114–30

[71] Aljumaily M M, Alsaadi M A, Binti Hashim N A, Mjalli F S, Alsalhy Q F, Khan A L and Al-Harrasi A 2020 Superhydrophobic nanocarbon-based membrane with antibacterial characteristics *Biotechnol. Progr.* **36** e2963

[72] Orion V and Imogen P 2024 Application of multi-objective optimization in controlling carbon footprint and costs in manufacturing *Front. Manage. Sci.* **3** 39–51

[73] Ghelich R, Jahannama M R, Abdizadeh H, Torknik F S and Vaezi M R 2019 Central composite design (CCD)-Response surface methodology (RSM) of effective electrospinning parameters on PVP-B-Hf hybrid nanofibrous composites for synthesis of HfB_2-based composite nanofibers *Compos. B Eng.* **166** 527–41

[74] Allaedini G, Tasirin S M and Aminayi P 2016 Yield optimization of nanocarbons prepared via chemical vapor decomposition of carbon dioxide using response surface methodology *Diam. Relat. Mater.* **66** 196–205

[75] Link H and Weuster-Botz D 2006 Genetic algorithm for multi-objective experimental optimization *Bioprocess. Biosyst. Eng.* **29** 385–90

[76] Khoshraftar Z, Ghaemi A and Hemmati A 2024 Comprehensive investigation of isotherm, RSM, and ANN modeling of CO_2 capture by multi-walled carbon nanotube *Sci. Rep.* **14** 5130

[77] Güler Ö, Tekeli M, Taşkın M, Güler S H and Yahia I 2021 The production of graphene by direct liquid phase exfoliation of graphite at moderate sonication power by using low boiling liquid media: the effect of liquid media on yield and optimization *Ceram. Int.* **47** 521–33

[78] Rehan H 2021 Energy efficiency in smart factories: leveraging IoT, AI, and cloud computing for sustainable manufacturing *J. Comput. Intell. Robot.* **1** 18

[79] Barati A, Shamsipur M, Arkan E, Hosseinzadeh L and Abdollahi H 2015 Synthesis of biocompatible and highly photoluminescent nitrogen doped carbon dots from lime: analytical applications and optimization using response surface methodology *Mater. Sci. Eng.* C **47** 325–32

[80] Krzywanski J, Sosnowski M, Grabowska K, Zylka A, Lasek L and Kijo-Kleczkowska A 2024 Advanced computational methods for modeling, prediction and optimization—a review *Materials* **17** 3521

[81] Shen S C, Khare E, Lee N A, Saad M K, Kaplan D L and Buehler M J 2023 Computational design and manufacturing of sustainable materials through first-principles and materiomics *Chem. Rev.* **123** 2242–75

[82] Sreejith S, Ajayan J, Uma Reddy N, Radhika J, Mathew J K, Sivasankari B and Raghavendra Reddy N 2024 Fundamentals of computational design in nanomaterials *Advanced Nanomaterials for Energy Storage Devices* (Berlin: Springer) 25–40

[83] Pistikopoulos E N and Tian Y 2024 Advanced modeling and optimization strategies for process synthesis *Annu. Rev. Chem. Biomol. Eng.* **15** 81–103

[84] Stavropoulos P and Panagiotopoulou V C 2022 Carbon footprint of manufacturing processes: conventional vs. non-conventional *Processes* **10** 1858

[85] Shahzad K, Mardare A I and Hassel A W 2024 Accelerating materials discovery: combinatorial synthesis, high-throughput characterization, and computational advances *Sci. Technol. Adv. Material, Meth.* **4** 2292486

[86] Tau O, Lovergine N and Prete P 2023 Adsorption and decomposition steps on Cu (111) of liquid aromatic hydrocarbon precursors for low-temperature CVD of graphene: a DFT study *Carbon* **206** 142–9

[87] Talayero C, Aït-Salem O, Gallego P, Páez-Pavón A, Merodio-Perea R G and Lado-Touriño I 2021 Computational prediction and experimental values of mechanical properties of carbon nanotube reinforced cement *Nanomaterials* **11** 2997

[88] Halim A F, Poinern G E J, Fawcett D, Sharma R and Surendran S 2025 Biomass-derived carbon nanomaterials: synthesis and applications in textile wastewater treatment, sensors, energy storage, and conversion technologies *CleanMat* **2** 4–58

[89] Abdelbasir S M, McCourt K M, Lee C M and Vanegas D C 2020 Waste-derived nanoparticles: synthesis approaches, environmental applications, and sustainability considerations *Front. Chem.* **8** 782

[90] Habib U, Ahmad F, Awais M, Naz N, Aslam M, Urooj M, Moqeem A, Tahseen H, Waqar A and Sajid M 2023 Sustainable catalysis: navigating challenges and embracing opportunities for a greener future *J. Chem. Environ* **2** 14–53

[91] El-Azazy M, Osman A I, Nasr M, Ibrahim Y, Al-Hashimi N, Al-Saad K, Al-Ghouti M A, Shibl M F, Al-Muhtaseb A A H and Rooney D W 2024 The interface of machine learning

and carbon quantum dots: from coordinated innovative synthesis to practical application in water control and electrochemistry *Coord. Chem. Rev.* **517** 215976

[92] Yadav P K, Chandra S, Kumar V, Kumar D and Hasan S H 2023 Carbon quantum dots: synthesis, structure, properties, and catalytic applications for organic synthesis *Catalysts* **13** 422

[93] Zhou Y, Wang C H, Lu W and Dai L 2020 Recent advances in fiber-shaped supercapacitors and lithium-ion batteries *Adv. Mater.* **32** 1902779

[94] Rajeshkumar L, Ramesh M, Bhuvaneswari V and Balaji D 2023 Carbon nano-materials (CNMs) derived from biomass for energy storage applications: a review *Carbon Lett.* **33** 661–90

[95] Zafar M, Imran S M, Iqbal I, Azeem M, Chaudhary S, Ahmad S and Kim W Y 2024 Graphene-based polymer nanocomposites for energy applications: recent advancements and future prospects *Results Phys.* **60** 107655

[96] Zhong M, Zhang M and Li X 2022 Carbon nanomaterials and their composites for supercapacitors *Carbon Energy* **4** 950–85

[97] Nyangiwe N N 2025 Applications of density functional theory and machine learning in nanomaterials: a review *Next Mater.* **8** 100683

[98] Gohar O, Khan M Z, Bibi I, Bashir N, Tariq U, Bakhtiar M, Karim M R A, Ali F, Hanif M B and Motola M 2024 Nanomaterials for advanced energy applications: recent advancements and future trends *Mater. Des.* **241** 112930

[99] Etesami M, Nguyen M T, Yonezawa T, Tuantranont A, Somwangthanaroj A and Kheawhom S 2022 3D carbon nanotubes-graphene hybrids for energy conversion and storage applications *Chem. Eng. J.* **446** 137190

[100] Hasani A, Teklagne M A, Do H H, Hong S H, Van Le Q, Ahn S H and Kim S Y 2020 Graphene-based catalysts for electrochemical carbon dioxide reduction *Carbon Energy* **2** 158–75

[101] Ugwu L I, Morgan Y and Ibrahim H 2022 Application of density functional theory and machine learning in heterogenous-based catalytic reactions for hydrogen production *Int. J. Hydrogen Energy* **47** 2245–67

[102] Savla N, Guin M, Pandit S, Malik H, Khilari S, Mathuriya A S, Gupta P K, Thapa B S, Bobba R and Jung S P 2022 Recent advancements in the cathodic catalyst for the hydrogen evolution reaction in microbial electrolytic cells *Int. J. Hydrogen Energy* **47** 15333–56

[103] Gusain R, Kumar N and Ray S S 2020 Recent advances in carbon nanomaterial-based adsorbents for water purification *Coord. Chem. Rev.* **405** 213111

[104] Jayaprakash N, Elumalai K, Manickam S, Bakthavatchalam G and Tamilselvan P 2024 Carbon nanomaterials: revolutionizing biomedical applications with promising potential *Nano Mater. Sci.*

[105] Adeola A O, Clermont-Paquette A, Piekny A and Naccache R 2024 Advances in the design and use of carbon dots for analytical and biomedical applications *Nanotechnology* **35** 012001

[106] Gayathri K and Vidya R 2024 Carbon nanomaterials as carriers for the anti-cancer drug doxorubicin: a review on theoretical and experimental studies *Nanoscale Adv.* **6** 3992–4014

[107] Terzapulo X, Kassenova A, Loskutova A and Bukasov R 2025 Carbon dots: review of recent applications and perspectives in bio-sensing and biomarker detection *Sens. Bio-Sens. Res.* **47** 100771

[108] Clermont-Paquette A, Larocque K, Piekny A and Naccache R 2024 Shining a light on cells: amine-passivated fluorescent carbon dots as bioimaging nanoprobes *Mater. Adv.* **5** 3662–74

[109] Hussain N, Bilal M and Iqbal H M 2022 Carbon-based nanomaterials with multipurpose attributes for water treatment: greening the 21st-century nanostructure materials deployment *Biomater. Polym. Horiz.* **1** 48–58

[110] Arora B and Attri P 2020 Carbon nanotubes (CNTs): a potential nanomaterial for water purification *J. Compos. Sci.* **4** 135

[111] Hsu C-Y, Rheima A M, Mohammed M S, Kadhim M M, Mohammed S H, Abbas F H, Abed Z T, Mahdi Z M, Abbas Z S and Hachim S K 2023 Application of carbon nanotubes and graphene-based nanoadsorbents in water treatment *BioNanoScience* **13** 1418–36

[112] Lazarenko N S, Golovakhin V V, Shestakov A A, Lapekin N I and Bannov A G 2022 Recent advances on membranes for water purification based on carbon nanomaterials *Membranes* **12** 915

[113] Adeola A O, Fuoco G, Adegoke K A, Adeleke O, Oyebamiji A K, Paramo L and Naccache R 2025 Experimental, machine-learning, and computational studies of the sequestration of pharmaceutical mixtures using lignin-derived magnetic activated carbon *ACS Sustain. Res. Manage.* **2** 219–30

[114] Vyas T, Parsai K, Dhingra I and Joshi A 2023 Nanosensors for detection of volatile organic compounds *Advances in Smart Nanomaterials and their Applications* (Amsterdam: Elsevier) 273–96

[115] Olawade D B, Fapohunda O, Usman S O, Akintayo A, Ige A O, Adekunle Y A and Adeola A O 2025 Artificial intelligence in computational and materials chemistry: prospects and limitations *Chem. Africa*

[116] Olawade D B, Ige A O, Olaremu A G, Ijiwade J O and Adeola A O 2024 The synergy of artificial intelligence and nanotechnology towards advancing innovation and sustainability—a mini-review *Nano Trends* **8** 100052

[117] Liu Y, Zhao T, Ju W and Shi S 2017 Materials discovery and design using machine learning *J. Materiomics* **3** 159–77

[118] Wang L, Tricard N, Chen Z and Deng S 2025 Progress in computational methods and mechanistic insights on the growth of carbon nanotubes *Nanoscale* **17** 11812–63

[119] Manan A, Zhang P, Ahmad S, Umar M and Raza A 2024 Machine learning prediction model integrating experimental study for compressive strength of carbon-nanotubes composites *J. Eng. Res.*

[120] Clinton L, Cubitt T, Flynn B, Gambetta F M, Klassen J, Montanaro A, Piddock S, Santos R A and Sheridan E 2024 Towards near-term quantum simulation of materials *Nat. Commun.* **15** 211

[121] Weidman J D, Sajjan M, Mikolas C, Stewart Z J, Pollanen J, Kais S and Wilson A K 2024 Quantum computing and chemistry *Cell Rep. Phys. Sci.* **5** 102105

[122] Mehta J, Dilbaghi N, Deep A, Hai F I, Hassan A A, Kaushik A and Kumar S 2025 Plastic waste upcycling into carbon nanomaterials in circular economy: synthesis, applications, and environmental aspects *Carbon* **234** 119969

[123] Haq F, Kiran M, Khan I A, Mehmood S, Aziz T and Haroon M 2025 Exploring the pathways to sustainability: a comprehensive review of biodegradable plastics in the circular economy *Mater. Today Sustain.* **29** 101067

[124] Gavankar S, Suh S and Keller A F 2012 Life cycle assessment at nanoscale: review and recommendations *Int. J. Life Cycle Assess.* **17** 295–303

Chapter 5

Nanoremediation of chemical pollutants using green carbon nanomaterials

Stephen Sunday Emmanuel and Ademidun Adeola Adesibikan

In the search for eco-efficient materials for environmental remediation, green carbon nanomaterials (GCNMs) have garnered the attention of researchers. This chapter focuses on the remediation of (in)organic pollutants using GCNMs. The goal is to analyze research findings, identify trends, and underscore intriguing future research hotspots. The efficiency of different GCNMs in the removal of various pollutants is critically evaluated. To shed light on potential eco-economic benefits, GCNMs' cyclic stability and reusability are also evaluated. The impact of coexisting pollutants is discussed to assess the feasibility of GCNMs in practical and industrial applications. Finally, challenges, research gaps, and potential avenues for further study are addressed.

5.1 Introduction

It is well known that there is growing concern about the presence of inorganic and organic pollutants [1, 2] in the environment, including water systems [3–5]. These pollutants mostly originate from anthropogenic sources, and their release into the environment without treatment has resulted in the degradation of ecosystem integrity [6, 7] and caused various health-related issues, including cancer, skin irritation, and endocrine disruption [8–10]. It is intriguing to note that one of the major channels through which these pollutants (radio-contaminants [11, 12], heavy metals [13], dyes [14–16], pharmaceutical and personal care products [17, 18], and pesticides [19]) reach humans, wreaking health havoc and disrupting fundamental ecological networks, is water [20, 21]. This is because water is the lifeline of nature, and all known life on Earth is heavily reliant on it [22].

Notably, many materials have been employed for the remediation of these pollutants. However, some of these materials face limitations such as regenerability/reusability, cost-effectiveness, and the use of toxic chemicals [23, 24].

Interestingly, GCNMs, regarded as a 'black knight for dirty waters,' have emerged as some of the most promising eco-sustainable materials in recent environmental cleanup research. These materials are eco-friendly, have good surface functionalities, are easy to synthesize, and their precursors are readily available; thus, they have garnered the attention of research communities [22, 24, 25]. Moreover, GCNMs combine the unique properties of carbon nanomaterials with the principles of green chemistry and sustainable development.

A thorough literature survey has revealed that numerous original research studies have been conducted on the use of GCNMs for the remediation of various organic and inorganic pollutants. Thus, there is a need for an update on the body of knowledge on this subject. For this purpose, this chapter evaluates the performance of GCNMs in the remediation of (in)organic pollutants while analyzing research findings and identifying trends. Furthermore, to bring potential eco-economic benefits to light, GCNMs' cyclic stability and reusability, as well as applications in real effluent, are also evaluated. Finally, salient dilemmas, challenges, frontiers, and intriguing future research hotspots are underscored.

5.2 Remediation of dye pollutants

Dyes are among the most popular organic compounds. They exemplify a double-edged sword in this era of industrialization, being indispensable in the textile, paper, polymer, and food sectors, yet responsible for environmental pollution and ecosystem disruption/disintegration [16, 24]. More specifically, the presence of dyes in water bodies has been linked to various ailments [14, 26]. Therefore, the remediation of these organic pollutants requires continuous attention. One of the methods recently explored for the removal of these harmful dye pollutants is the use of GCNMs due to their eco-friendliness and efficiency. For example, green N-doped carbon quantum dots (CQDs) were fabricated from grass and explored for the removal of cationic and anionic dye pollutants through photocatalytic technology. Notably, the findings showed that the GCNM delivered 100% remediation efficiency for all the dye pollutants (acid blue, methyl orange (MO), eosin yellow, methylene blue (MB), Eriochrome black T, and acid red) investigated. However, while other dyes took 30 min to degrade completely, MO and MB took 90 min. This slightly longer duration was suspected to be due to the complex nature of MO and MB, including the presence of the phenothiazine ring and the azo bond, which take longer for the radicals to break [27]. In a similar study, when prickly-pear-derived carbon dots (CDs) were employed for the removal of basic fuchsin, methyl green, and crystal violet (CV), an outstanding removal efficiency of 90% was achieved for the three dye pollutants after 60 min at pH 7.5 and 20 °C [28].

In another study [29], CQDs were prepared from *Aloe vera* and examined for the removal of eosin yellow with the assistance of light irradiation. According to the findings, the green CQDs had outstanding photocatalytic efficiency, reaching 98.55% removal in 80 min and 100% in 100 min. It was opined that the blue emission and homogeneous dispersion of the GCNM in an aqueous solution, which enhance efficient dye degradation, were facilitated by the presence of -OH groups [29]. Comparable efficiencies were also recorded for palm kernel shell, activated carbon(AC)@graphene oxide(GO)@iron oxide [30], and areca nut husk, GO@MnO@NiO@ZnO [31], for the remediation of Acid Blue 11 and MB, respectively.

In another experiment, a green carbon nanocomposite made from orange peel, CDs@ZnO, was used for the remediation of naphthol blue black dye. Interestingly, within 45 min of reaction, the GCNM yielded a remediation efficiency of roughly 100%, while when only ZnO was used, a dye removal efficiency of only 83% was recorded under the same experimental conditions, including light irradiation [32]. This showcases the contribution of the orange peel CDs to the performance of the nanocomposite. Mallikarjuna *et al* supported this result in their study, where it was found that AgCDs@g-C$_3$N$_4$ gave a better remediation performance (97%) for the removal of RhB in 25 min compared to bare g-C$_3$N$_4$ (59%) under the same conditions. The loading of Ag quantum dots (QDs) into g-C$_3$N$_4$ allowed the restriction, recombination, and migration of photoproduced e-/h$^+$ pairs, resulting in an improvement in dye degradation performance [33].

Another research team [34] studied the potential of rice-straw-derived GO nanoplatelets for the removal of CV. The results showed that the GCNM had an excellent adsorption capacity for the cationic dye pollutant, with a dye removal efficiency of 99.73% after 120 min at pH 8 [34]. This is comparable to the remediation performances of sugarcane bagasse reduced graphene oxide (rGO) for MB in 30 min [35], WO$_3$@N-sugarcane bagasse CQDs for MB in 180 min [36], and peanut shell biochar (BC)@GO for CV/bromophenol blue in 60 min, as presented in table 5.1 [37]. Similarly, a study was conducted on the capacity of laccase carbon nanotubes (CNTs) for the elimination of Congo red (CR) dye. The research findings showed that at pH 7, 96% of the anionic azo dye was removed in 180 min [38]. This is consistent with the remediation performance recorded for multiwalled carbon nanotubes (MWCNTs)@TiO$_2$@chitosan for the removal of malachite green (MG) [39] and CNTs@*Saccharum munja* plant biomass for the removal of Safranin O and MB [40] at pH 7. Other important studies that have examined the remediation of dye pollutants using GCNMs are summarized in table 5.1, and the data show that GCNMs have a similar pattern of remarkable efficiency over a broad spectrum of pH values (3–10) in shortest times ranging from 10 to 240 min, with nanocomposites coming to the fore.

5.3 Remediation of pharmaceutical and personal care products

Pharmaceutical and personal care products are another class of organic compounds that are a boon to human and animal health but a bane to the environment, where they are unwanted and thus become a serious pollutant of concern. Beyond using mere carbon materials, GCNMs have proven to be effective in the remediation of PPCPs, especially from aquatic systems. For example, a rice-straw-derived fluorescent CDs@polydopamine imprinted polymer was employed for the aqueous remediation of ibuprofen. The results revealed that this GCNM demonstrates strong sensitivity (a limit of detection (LOD) of 1.58×10^{-5} µM) and selectivity toward ibuprofen in the presence of other pharmaceutical pollutants such as norfloxacin,

Table 5.1. Summary of the performance of GCNMs in the removal of dye pollutants.

GCNM	Surface area (m^2 g^{-1})	Dye pollutant	Removal efficiency (%)	Time (min)	pH	Reusability (% removal in 1st cycle–% removal in (nth cycle))	Reference
Orange peel and banana peel AC@chitosan	480	MB	93.10	60	7	98.9–94.3 (4th)	[41]
MWCNT@TiO$_2$@chitosan	198.2	MG	98.53	15	7	97.65–>70 (7th)	[39]
Laccase CNTs	119.4	CR	96	180	7	—	[38]
Biomass-derived carbon@montmorillonite clay	29	MB	99	180	—	—	[42]
Graphene@CNTs	78.9	MB	97	120	—	>85–~70 (4th)	[43]
MWCNT@sugarcane bagasse BC	390	MB	64	~120	7	—	[44]
MWCNT@hickory chip BC	351	MB	47	~120	7	—	[44]
CNT@S. munja plant biomass	—	Safranin O	94.87	10	7	>98–>95 (10th)	[40]
CNT@S. munja plant biomass	—	MB	92.68	10	7	>97–>94 (10th)	[40]
MWCNT@orange peel	—	MB	70	—	10	—	[45]
CNT@pearl millet stem	—	Safranin O	94	10	7	98.81–94.45 (10th)	[46]
CNT@pearl millet stem	—	MB	93	10	7	98.11–93.45 (10th)	[46]
Magnetic sugarcane bagasse AC@CNT	109.07	MB	94.03	120	7	100–>80 (4th)	[47]
CNT@eucalyptus AC	101	Eosin yellow	>90	<60	—	—	[48]
CNT@eucalyptus AC	101	MB	>90	<30	—	—	[48]
CNT@*Mucor circinelloides*@Fe$_2$O$_3$	—	CR	85	—	—	—	[49]
N-TiO$_2$@waste biomass CNT	—	CR	99.90	20	—	—	[50]
TiO$_2$@waste biomass CNT	—	CR	99.90	20	—	—	[50]
WO$_3$@N-sugarcane bagasse CQDs	25.3	MB	96.86	180	—	>90–86.85 (3rd)	[36]
AgQDs@g-C$_3$N$_4$	—	Rhodamine B (RhB)	97	25	—	—	[33]
ZnO@coffee grounds CDs	—	MB	80	150	—	—	[51]
ZnO@CD	—	MG	94.8	60	8	94.8–75 (4th)	[52]

Sustainable Carbon Nanomaterials and their Applications

CoFe$_2$O$_4$@turmeric CQDs	Acid black	—	95	60	—	—	[53]
CoFe$_2$O$_4$@turmeric CQDs	Acid brown	—	90	90	—	—	[53]
CoFe$_2$O$_4$@turmeric CQDs	Acid red	—	65	120	—	—	[53]
Polyalthia longifolia CQDs	MB	—	>70	60	8	—	[54]
Polyalthia longifolia CQDs	CR	—	>70	90	6	—	[54]
Orange peel CDs@ZnO	Naphthol blue black	—	100	45	—	—	[32]
Hyacinth CDs	Skycion yellow HE 4 R	—	70.67	—	—	—	[55]
Aloe vera CQDs	Eosin yellow	—	100	100	—	—	[29]
Prickly pear CDs	CV	—	90	60	7.5	—	[28]
Prickly pear CDs	Basic fuchsin	—	90	60	7.5	—	[28]
Prickly pear CDs	Methyl green	—	90	60	7.5	—	[28]
Potato peel CDs@Fe$_3$O$_4$	RhB	—	70	240	—	—	[56]
Potato peel CDs@Fe$_3$O$_4$	MG	—	80	120	—	—	[56]
Grass N-CQDs	Acid blue	—	100	30	—	—	[27]
Grass N-CQDs	Acid red	—	100	30	—	—	[27]
Grass N-CQDs	Eosin yellow	—	100	90	—	—	[27]
Grass N-CQDs	MB	—	100	90	—	—	[27]
Grass N-CQDs	MO	—	100	60	—	—	[27]
Neem bark CQDs	MB	280.39	64.7	—	7	—	[57]
Palm kernel shell AC@GO@iron oxide	Acid blue 113	—	96.30	—	7	99.8–74.0 (5th)	[30]
Rice husk LaFeO$_3$@GO	MB	72.27	99.3	120	8	—	[58]
Rice husk LaFeO$_3$@GO	CR	72.27	98.0	120	3	—	[58]
Rice straw GO	CV	—	99.73	120	8	90.74 (8th)	[34]
Areca nut husk	MB	18.263	93.05	180	—	—	[31]
GO@MnO@NiO@ZnO							

(Continued)

Table 5.1. (*Continued*)

GCNM	Surface area (m² g⁻¹)	Dye pollutant	Removal efficiency (%)	Time (min)	pH	Reusability (% removal in 1st cycle–% removal in (nth cycle))	Reference
Areca nut husk GO	142.369	MB	90.76	180	—	—	[31]
Sawdust BC@GO@LaO	458.92	Indigo carmine	>70	120	4	—	[59]
Sugarcane bagasse rGO	—	MB	92	30	—	>90–47.2 (3rd)	[35]
GO@aminated lignin aerogels	16.11	MG	91.72	—	8	>90–89.8 (5th)	[60]
Peanut shell BC@GO	—	CV	97	60	8	98–63(5th)	[37]
Peanut shell BC@GO	—	Bromophenol blue	87	60	6	87–51 (5th)	[37]
Pomegranate peel waste BC@MgAl-layered double hydroxide (LDH)	275.57	MB	98.54	25	7	97.6–82.7 (5th)	[61]

ketoprofen, levofloxacin, and aspirin. Furthermore, adsorption experiments confirmed that the GCNM achieved a remediation efficiency of 99.9%, which is equivalent to an adsorption capacity of 209.8 mg g^{-1} in 120 min at pH 7. This outcome revealed the dual function of the rice-straw-derived fluorescent CDs@polydopamine imprinted polymer, namely the simultaneous detection and cleanup of pharmaceutical pollutants [62]. A similar efficiency was achieved when blue-algae-fabricated CQDs@TiO$_2$ were used to eliminate Favipiravir, an antiviral drug that gained momentum during the COVID-19 outbreak. Specifically, 98.6% of this pharmaceutical pollutant was removed in 60 min, which is roughly 16× greater than the removal rate of a typical organic-derived material, CQDs@TiO$_2$ [63], and this result is comparable with the recorded results (~100%) for the removal of losartan, ofloxacin, sodium diclofenac, and ciprofloxacin using sugarcane bagasse MWCNTs and eucalyptus leaf MWCNTs [64]. The authors of [63] explained that the copious C=O, P–O, OH, and C–O–C functionality and small particle size (1.42 nm) of the green CQDs, combined with the enhanced TiO$_2$ quantity, were the key reasons for the enhanced interaction of the blue algae CQDs@TiO$_2$ with the pharmaceutical pollutant, which resulted in outstanding remediation [63]. On the other hand, when sugarcane bagasse MWCNTs were used to remove propranolol, a lower efficiency of 90% was achieved. This small difference is due to the fact that propranolol pollutant molecules are more stable and thus slightly tougher to eliminate from aqueous solutions [64].

In another study [65], sewage sludge CQDs@Bi$_2$MoO$_6$@Bi$_2$S$_3$ with a surface area of 59.5 m^2 g^{-1} were used to remove four pharmaceutical pollutants. Notably, except for ciprofloxacin, where only 80.6% was eliminated, over 90% of oxytetracycline, doxycycline, and tetracycline hydrochloride were eliminated in 120 min at pH 4.5. This percentage removal was higher than the recorded results for Bi$_2$MoO$_6$@Bi$_2$S$_3$, demonstrating the impact of the green CQDs in CQDs@Bi$_2$MoO$_6$@Bi$_2$S$_3$. This is because the green CQDs can enhance the generation of ·O$_2^-$ and ·OH, which are important radicals in the remediation of pharmaceutical pollutants. A similar remediation efficiency was also achieved by withered peach CQDs@Bi$_2$WO$_6$@Cu$_2$O [66] and withered magnolia CDs@Bi$_2$WO$_6$ [67] for the removal of tetracycline (TC) in 90 min. Additionally, the removal efficiency of the green CD composites was superior to that of their bare nanoparticle counterparts (Bi$_2$WO$_6$ and Bi$_2$WO$_6$@Cu$_2$O). While the green CDs enhance the surface functional groups of the GCNMs, the heterojunction of Bi$_2$WO$_6$/Cu$_2$O boosts the separation of e-/h$^+$, which is an essential process in the remediation of pharmaceutical pollutants.

Moreover, plantain peel rGO@Zn@Cu was explored for the removal of ibuprofen and diclofenac. The results revealed that this GCNM was able to eliminate over 90%, which is equivalent to 148 mg g^{-1} and 146 mg g^{-1}, respectively. Kinetically, the remediation process was best described by the pseudo-second-order model, which shows that the pharmaceutical pollutants were largely removed via chemisorption [68]. Similarly, when MWCNTs-OH/COOH-Fe@coffee were employed as a GCNM for the removal of diclofenac and losartan, the remediation efficiency was greater than 90%, as presented in table 5.2. However, when the elimination rates of the two pharmaceuticals are compared, it is clear that diclofenac

Table 5.2. Summary of the performance of GCNMs in the removal of pharmaceutical and personal care products.

GCNM	Surface area (m² g⁻¹)	Pollutant	Removal efficiency (%)	Time taken (min.)	pH	Reusability (% removal in 1st cycle–% removal in (nth cycle))	Reference
Gliricidia sepium BC	28	Oxytetracycline	>70		6		[70]
Withered magnolia CDs@Bi$_2$WO$_6$	23	TC	80	90	—	74–>60 (5th)	[67]
Withered peach CQDs@Bi$_2$WO$_6$@Cu$_2$O	—	TC	80	90	—	—	[66]
Plantain peel-rGO@Zn@Cu	—	Ibuprofen	>90	—	—	80 (4th)	[68]
Plantain peel-rGO@Zn@Cu	—	Diclofenac	>90	—	—	80 (4th)	[68]
MWCNTs-OH/COOH-Fe@coffee	121.232	Losartan	98.26	20	—	78.64–66.69 (3rd)	[69]
MWCNTs-OH/COOH-Fe@coffee	121.232	Diclofenac	98.42	5	—	86.12–60.68 (3rd)	[69]
Rice straw CDs@polydopamine imprinted polymers	184.09	Ibuprofen	99.90	120	7	—	[62]
Sewage sludge CQDs@Bi$_2$MoO$_6$@Bi$_2$S$_3$	59.50	TC hydrochloride	95.9	120	4.5	>85–>70 (5th)	[65]
Sewage sludge CQDs@Bi$_2$MoO$_6$@Bi$_2$S$_3$	59.50	Doxycycline	91.6	120	4.5	—	[65]
Sewage sludge CQDs@Bi$_2$MoO$_6$@Bi$_2$S$_3$	59.50	Oxytetracycline	92.5	120	4.5	—	[65]
Sewage sludge CQDs@Bi$_2$MoO$_6$@Bi$_2$S$_3$	59.50	Ciprofloxacin	80.6	120	4.5	—	[65]
Blue algae CQDs@TiO$_2$	—	Favipiravir	98.6	60	—	~98–94.6 (5th)	[63]
Sugarcane bagasse MWCNTs	—	Ciprofloxacin	~100	—	—	—	[64]

Sugarcane bagasse MWCNTs	Sodium diclofenac	~100	—	—	—	[64]
Sugarcane bagasse MWCNTs	Losartan	~100	—	—	—	[64]
Eucalyptus leaves MWCNTs	Losartan	~100	—	—	—	[64]
Eucalyptus leaves MWCNTs	Sodium diclofenac	~100	—	—	—	[64]
Eucalyptus leaves MWCNTs	Ofloxacin	~100	—	—	—	[64]
Eucalyptus leaves MWCNTs	Ciprofloxacin	~100	—	—	—	[64]
Sugarcane bagasse MWCNTs	Propranolol	90	—	—	—	[64]

elimination was more efficient than losartan elimination, suggesting that the latter has a lower affinity for the GCNM. Additionally, the diclofenac elimination mechanism has a much quicker equilibrium time (5 min) compared to that of losartan (20 min). The molecular structure of the compounds may be the reason for these differences, as the electrical and molecular characteristics of diclofenac molecules are more reactive than those of losartan, promoting their elimination [69].

5.4 Remediation of agrochemicals and phenols

Agrochemicals and phenols are another class of organic pollutants that have long been considered a necessary evil, vital for manufacturing and agricultural purposes, yet increasingly scrutinized for their severe health impacts on humans and the ecosystem at large. Thus, their removal from the environment, where they are unwanted, is important. As presented in table 5.3, efforts have been made to employ GCNMs to remove various agrochemicals and phenols from the environment, especially aquatic systems. For instance, in a particular study, water hyacinth leaves were employed to fabricate a green CD composite with g-C_3N_4. This GCNM was used to photocatalytically remove 2,4-dichlorophenol. Remarkably, in 120 min, 94% of the phenolic pollutant was degraded by the GCNM containing the highest quantity of CDs, mainly through the actions of h^+ and O_2^- [71]. A similar efficiency (90%) was also reported for the removal of Bisphenol A in 120 min using rice straw CQDs@BiOCl [72], which is comparable to the results obtained for the cleanup of Phoxim and Malathion in 120 min using activated citrus peel@GO@Fe_3O_4 [73].

In another report [76], CDs synthesized from *Ficus benghalensis* were incorporated into mesoporous silica to obtain a GCNM, which was used to remove Methoxy-DDT. At an acidic pH of 2.21, over 80 % of the endocrine-disrupting pollutant was removed in just 300 min. This is consistent with another study, where graphene@algal biomass was used to remove over 70% of 2,4-dichlorophenoxy-acetic acid in just 90 min at pH 2.0 and 30 °C [74].

In another study [75], olive pomace GQDs@Fe_3O_4 were used to remove 87.19% of Malathion in 210 min at pH 7 and 25 °C. Interestingly, when the temperature was increased to 50 °C, the amount of Malathion removed increased to 97.24% in the same 210 min at pH 7 [75]. This shows the potential of the GCNM to work effectively if employed in a harsh or hot real-world environment. In 210 min, comparable remediation efficiency was also reported for the removal of Bisphenol A using *Colocasia esculenta* leaf reduced CDs under light irradiation [79], which is congruent with the results obtained when sugarcane bagasse MWCNTs [64] and eggshell carbon@Fe_3O_4 [78] were employed for the removal of Bisphenol A and 2,4,6-trichlorophenol.

5.5 Remediation of heavy metals and radioactive ions

Heavy metal and radioactive ions have been a prominent class of inorganic pollutants since time immemorial, owing to continuous technological advancement [11, 22]. Thus, their remediation requires persistent effort, including sourcing eco-friendly materials such as GCNMs. For example, green graphene-like materials

Sustainable Carbon Nanomaterials and their Applications

Table 5.3. Summary of the performance of GCNMs in the removal of agrochemicals and phenols.

GCNM	Surface area (m^2 g^{-1})	Agrochemical or phenol	Removal efficiency (%)	Time taken (min)	pH	Reusability (% removal in 1st cycle–% removal in (nth cycle))	Reference
Graphene@algal biomass	—	2,4-Dichlorophenol (2,4-D)	>70	90	2	—	[74]
Hyacinth leaf CDs@g-C_3N_4	159	2,4-D	94	120	—	—	[71]
Rice straw CQDs@BiOCl	—	Bisphenol A	90	120	—	84–66 (5th)	[72]
Olive pomace graphene quantum dots (GQDs)@Fe_3O_4	22.10	Malathion	97.24	210	7	79.74–70.64 (3rd)	[75]
Activated citrus peel@GO@Fe_3O_4	—	Phoxim	>80	120	—	~99–>85 (5th)	[73]
Activated citrus peel@GO@Fe_3O_4	—	Malathion	>80	120	—	~100–>85 (5th)	[73]
Ficus benghalensis CDs@silica	185.539	Methoxy-DDT	>80	300	2.21	86.53 (5th)	[76]
Sugarcane bagasse MWCNTs	—	Bisphenol A	~100	—	—	—	[64]
Magnetic carbon nanofiber	312.65	Bisphenol A	>70	—	—	100–96 (10th)	[77]
Eggshell carbon@Fe_3O_4	—	2,4,6-Trichlorophenol	~100	—	—	—	[78]
Eggshell carbon@Fe_3O_4	—	Bisphenol A	~100	—	—	—	[78]
Colocasia esculenta leaf reduced CDs	—	Bisphenol A	100	210	—	—	[79]

were fabricated from three different types of agro-wastes (sugarcane bagasse, sawdust, and coconut shell) and explored for the remediation of Cu(II). Interestingly, all three GCNMs showed excellent Cu(II) removal efficiency in less than 2 h, and their remediation operation was a good fit for the pseudo-second-order kinetic model, which suggests strong chemical interaction between the heavy metal pollutant and the GCNMs. However, sugarcane bagasse graphene with a surface area of 15.27 m^2 g^{-1} exhibited a far superior efficiency of 95% in 30 min, in contrast to coconut shell graphene (55.98 m^2 g^{-1}) and sawdust graphene (24.15 m^2 g^{-1}), which achieved 83% and 70% in 120 min, respectively, despite having better surface areas. The superiority of both sugarcane bagasse and coconut shell graphene-like GCNMs might be due to the fact that they have a morphological pattern of rGO rather than mere graphene [80], and this is in good agreement with results obtained for the removal of Pb(II) using oil palm sap rGO [81] and xanthan@GO@TiO$_2$ [82]. Similarly, in another comparative study, oil palm husk-derived GO and its composite with ZnO were explored for the bioremediation of Cd(II) in microbial fuel cells. It was revealed that the composite achieved a superior removal efficiency of 90% compared to green GO alone, which removed 83.50%. Even though the green GO had a superior surface area, the ZnO clearly had a positive impact on the performance of the composite [83]. This pattern of results is consistent with what was achieved (97%) for the removal of Cd(II) and Cr(VI) using acorn peel GO@Mg$_{1-x}$Ca$_x$Fe$_2$O$_4$ [84] and GO@fungal hyphae [85], but at different residence times.

Another study explored the efficiency of rice-husk-derived nanobiochar (NBC) for the remediation of Pb(II) and Cd(II) from aqueous media and further examined the optimized operating conditions. The study findings revealed that the NBC eliminated 96% of Pb at pH 6 and 91% of Cd at pH 8 in 360 min. Conversely, when compared to bare rice husk biosorbent, 90% of Pb and 87% of Cd were removed [86]. In addition, according to the authors, a continuous steady increase in the adsorption efficiency with residence time indicated that a multilayer remediation process was at work. A comparable remediation output (>97%) was also recorded for the removal of Cd, Cr, and Ni using water hyacinth and tea-waste-derived NBC, in contrast to bare BC, which exhibited lower efficiency under the same optimized experimental conditions [87]. This highlights the superiority of NBC as an efficient GCNM for the remediation of heavy metals.

In another study [88], N-doped CDs were synthesized from red *Malus floribunda* fruits and investigated for the aqueous removal of Cr(III), Pb(II), Cd(II), and Hg(II). Remarkably, all the heavy metals were effectively removed by the GCNMs in multiple systems, with Hg(II) showing the highest removal rate of 99%. This result was due to the various functional groups such as amines, hydroxyls, carboxyls, and carbonyls present in the GCNM, which allowed for excellent interactions with the heavy metal pollutants. Additionally, the high removal efficiency of Hg(II) can be explained by the hard–soft acid–base principle in that the nitrogen/amine functional moieties exhibit a strong attraction for the soft acid Hg(II) ions [88]. Also, when oil palm husk CDs and waste tobacco leaf CDs were employed for the remediation of Hg(II), an outstanding efficiency of 84.6% [89] and 99.40% [90] was recorded at an optimized pH of 7 and 6.8 in 30 min, respectively. Moreover, the fluorescence characteristics of waste tobacco-leaf-derived CDs allow for not only the removal of Hg(II) but also its detection, and this is an added advantage [90].

Uranium, a notorious radioactive pollutant, was also removed by sponge-like 3D Konjac glucomannan GO. The findings revealed that the highest adsorption capacity of sponge-like GCNM was 266.97 mg g^{-1} for U(VI), which was equivalent to an efficiency of more than 70% in 100 min at pH 5, and this was far higher than the removal rate achieved when only Konjac glucomannan was used. This again attests to the positive impact of GO, a very good carbon nanomaterial. Additionally, in a multi-ion environment, the GCNMs demonstrated outstanding selectivity for uranium trapping [91]. Similar remediation efficiency was recorded for polymeric CQDs at pH 5 for the elimination of U(VI), but in just 1 min. The remediation operation was a better fit for the pseudo-second order models, which indicated a chemisorption operation [92]. As presented in table 5.4, various GCNMs can eliminate different heavy metals over a wide range of pH values spanning from acidic to basic, and this suggests that these GCNMs can be employed in harsh environments. Overall, GCNMs show a similar heavy metal/radio-contaminant removal efficiency pattern through complexation, pore filling, ion exchange, cation–π interaction, and electrostatic attraction.

5.6 Remediation of other pollutants

As summarized in table 5.5, GCNMs have been employed for the remediation of other notable pollutants such as perfluorooctanoic acid (PFOA) [98], crude oil [99], glyphosate [70], and polycyclic aromatic hydrocarbons (naphthalene, fluoranthene, and anthracene) [100] using mussel shell CQDs@TiO$_2$, guava leaf CDs, *Gliricidia sepium* NBC, and chitosan@GO@Fe$_3$O$_4$. The performance of various GCNMs follows a similar trend, with a remediation rate of more than 70% within a pH range of 6.5 to 7.0. While this is commendable, especially in the case of oil spill cleanups, the fabrication of GCNMs with more specificity/affinity for the poorly adsorbed PFOA and fluoranthene is recommended for future research. This is because per- and polyfluoroalkyl substances (PFASs) and polycyclic aromatic hydrocarbons have complex, diverse characteristics; thus, a GCNM that works excellently for one pollutant might work poorly for another.

5.7 Recyclability and stability dynamics

Recyclability testing involves the study of the number of times a material can be reused in a remediation process for the removal of pollutants. This study is very important, as it shows how stable a material is and its potential for pilot-scale and industrial application [41, 43]. Moreover, recyclability and stability studies provide background knowledge about the eco-friendliness and cost-effectiveness of a material or process in general [40].

In summary, recycling the spent GCNM involves the following main steps. First, it is separated from the reaction mixture using techniques such as decantation, centrifugation, filtration, or magnetic separation. Second, the sorbed pollutants or

Table 5.4. Summary of the performance of GCNMs in the removal of heavy metals and radioactive pollutants.

GCNM	Surface area (m² g⁻¹)	Heavy metal	Removal efficiency (%)	Time taken (min.)	pH	Reusability (% removal in 1st cycle-% removal in (nth cycle))	References
Oil palm husk GO@ZnO	12.8	Cd(II)	90.00	—	—	—	[83]
Oil palm husk GO	280.1	Cd(II)	83.50	—	—	—	[83]
Sugarcane bagasse graphene	15.27	Cu(II)	95	30	—	—	[80]
Coconut shell graphene	55.98	Cu(II)	83	120	—	—	[80]
Sawdust graphene	24.15	Cu(II)	70	120	—	—	[80]
Water hyacinth NBC		Cd(II)	99.8	—	—	—	[87]
Tea waste NBC		Cd(II)	99.8	—	—	—	[87]
Water hyacinth NBC		Cr(VI)	98.8	—	—	—	[87]
Tea waste NBC		Cr(VI)	98.7	—	—	—	[87]
Water hyacinth NBC		Ni(II)	98.6	—	—	—	[87]
Tea waste NBC		Ni(II)	97.30	—	—	—	[87]
Gliricidia sepium NBC	28	Cr(VI)	>70	—	9	—	[70]
Gliricidia sepium NBC	28	Cd(II)	>70	—	10	—	[70]
Rice husk NBC		Pb(II)	96	360	6	—	[86]
Rice husk NBC		Cd(II)	91	360	8	—	[86]
Broccoli stem CDs@SA@PEI		Pb(II)	>90	360	—	85.67 (5th)	[93]
Tapioca flour CDs		Pb(II)	80.6	260	—	—	[94]
Red M. floribunda fruit CD hydrogel		Hg(II)	99	—	—	—	[88]
Red M. floribunda fruit CD hydrogel		Pb(II)	82	—	—	—	[88]
Red M. floribunda fruit CD hydrogel		Cd(II)	72	—	—	—	[88]
Red M. floribunda fruits CDs hydrogel		Cr(III)	80	—	—	—	[88]
Oil palm husk CDs		Hg(II)	84.6	30	7	—	[89]
Waste tobacco leaf CDs		Hg(II)	99.40	30	6.8	—	[90]
Oil palm trunk GO@PANI@Ag		Pb(II)	78.10	—	—	—	[95]

Oil palm trunk GO@PANI@Ag	—	Cd(II)	80.25	—	—	—	[95]
Shrimp waste chitosan@Schiff base ligand@GO	—	Cu(II)	>70	420	6	—	[96]
Shrimp waste chitosan@Schiff base ligand@GO	—	Pb(II)	>70	420	6	—	[96]
Oil palm sap rGO	—	Pb(II)	90	—	—	—	[81]
Xanthan@GO@TiO$_2$	—	Pb(II)	80.81	—	5.2	84.78 (5th)	[82]
Oak fruit peel GO@Mg$_{1-x}$Ca$_x$Fe$_2$O$_4$	11.56	Cd(II)	97	20	—	82 (3rd)	[84]
GO@fungal hyphae	—	Cr(VI)	97	540	2	97–80 (5th)	[85]
Potato peel waste AC	52	Pb(II)	84	60	—	—	[97]
CQDs@poly(anthranilic acid-formaldehyde-phthalic acid) (PAFP)	28.79	U(VI)	98.0	1	5	97.1–85.2 (5th)	[92]
Konjac glucomannan GO	—	U(VI)	>70	100	5	—	[91]
Sewage sludge CQDs@Bi$_2$MoO$_6$@Bi$_2$S$_3$	59.50	Cr(VI)	88.60	120	4.5	>85->70 (5th)	[65]

Table 5.5. Summary of the performance of GCNMs in the remediation of other pollutants.

GCNM	Surface area ($m^2\,g^{-1}$)	Pollutant	Removal efficiency (%)	Time taken (min.)	pH	Reusability (% removal in 1st cycle–% removal in (nth cycle))	Reference
Gliricidia sepium BC	28	Glyphosate	>70		7	—	[70]
Guava leaf CDs	—	Oil spill	98	—	—	91 (5th)	[99]
Mussel shell CQDs@TiO$_2$	—	PFOA	29	480	—	—	[98]
Chitosan@GO@Fe$_3$O$_4$	—	Naphthalene	93.55	30	6.5	—	[100]
Chitosan@GO@Fe$_3$O$_4$	—	Anthracene	89.25	30	6.5	—	[100]
Chitosan@GO@Fe$_3$O$_4$	—	Fluoranthene	62.28	30	6.5	—	[100]

degradation by-products are removed from the recovered material with the aid of an eluent. Finally, the spent GCNM is regenerated or reactivated before being employed in the next round of the remediation process [101–104]. Notably, many regeneration methods exist [105, 106], among which the chemical method is preferred owing to its fast regeneration, efficiency, affordability, and ease of operation [102]. For instance, the regeneration and recycling efficiency of broccoli stem CDs@SA@PEI were examined for the remediation of Pb(II). To achieve the regeneration of the GCNM, various acids were used to desorb Pb(II) via ion exchange; EDTA-2Na was used to complex with Pb(II); and an alkali was used to trigger Pb(OH)$_2$ precipitation. For a GCNM retaining around 96.72% of the initial adsorption capacity, the findings indicated that EDTA-2Na had the best elution output of all the eluents. Furthermore, the GCNM retained 85.7% of its initial adsorption capacity after five successive rounds [93]. This is consistent with the recyclability performance of xanthan@GO@TiO$_2$ [82], oak fruit peel GO@Mg$_{1-x}$Ca$_x$Fe$_2$O$_4$ [84], GO@fungal hyphae [85], and CQDs@poly(anthranilic acid-formaldehyde-phthalic acid) (PAFP) for the removal of Pb(II), Cd(II), Cr(VI), and U(VI) using 0.2 M HCl, 1 M NaOH, and 0.1 M HCl as eluents, respectively.

In another study [36], the stability of N-CQDs@WO$_3$ was investigated by testing their recyclability potential. In this study, 5 mg l^{-1} of MB dye pollutant was utilized for each experimental round. The GCNM was retrieved from the prior reaction mixture and continuously utilized without any cleansing. The result demonstrated that the GCNM has good stability and was able to eliminate up to 86.85% of the MB dye pollutant even after three rounds of recyclability [36]. This is comparable to the recyclability efficiency (83.70% at the 3rd cycle) exhibited by Gum Ghatti powder CDs@ZnO for the remediation of MG dye pollutant, where ethanol (EOH) and water were employed as eluents [52]. According to the authors of [52], the small reduction witnessed in the remediation efficiency could be due to the loss of GCNM during the centrifugation process. Similarly, at the eighth round of the recyclability study, rice straw-derived GO also achieved a remarkable CV dye pollutant cleanup efficiency of 90.74 [34]. This is in very good agreement with the recyclability performance of GO@aminated lignin aerogels for the removal of MG. After five application rounds, the adsorption efficiency decreased somewhat but was still over 89.8% [60]. On the one hand, the entire procedure resulted in a decline in the quality of the GO@aminated lignin aerogels. Conversely, a reduction in adsorption sites resulted from inadequate elution during the desorption process [50]. Furthermore, judging by its appearance and form, the GCNM maintained its initial hardness and shape even after several adsorption–desorption rounds. After several cycles, the GCNM structure did not exhibit any notable morphological deformation or disintegration, retaining its exceptional mechanical strength. This outcome was attributed to the GCNM's three-dimensional porous architecture and numerous hydrogen bonds, as well as the robust support provided by the aminated lignin [60].

Another study examined the recyclability of the removal of ibuprofen and diclofenac (pharmaceutical pollutants) by plantain peel-rGO@Zn@Cu. Following the remediation operation, the GCNM loaded with the pharmaceutical pollutants was regenerated with EOH, thoroughly cleaned with water, and then utilized again for further pharmaceutical pollutant removal. Four reuses of the GCNM were completed, and the results demonstrated that no discernible decrease in efficiency was seen over the four rounds [68]. A similar recyclability performance was also reported for blue algae CQDs@TiO$_2$ [63], activated citrus peel@GO@Fe$_3$O$_4$ [73], *Ficus benghalensis* CDs@silica [76], magnetic carbon nanofiber [77], and guava leaf CDs [99] for the remediation of Favipiravir, Phoxim/Malathion, Methoxy-DDT, Bisphenol A, and crude oil, respectively. In addition, structural and morphological characterizations revealed that rice-straw-derived CQDs@BiOCl had good stability even after being reused five times for the removal of Bisphenol A [72].

In another regenerability–recyclability study for the removal of Safranin O and MB dye by CNTs@*S. munja* plant biomass, different eluting agents (room-temperature water, hot water, HCl, and acetone) were tested. Interestingly, hot water and acetone gave the best regeneration/washing output. Thus, this eluent combination was used to successfully desorb Safranin O and MB from GCNM. It was discovered that following the tenth cycle, there was no discernible decrease in GCNM removal effectiveness (>98% in the first round to >94% in the tenth round), suggesting that this GCNM may find use on an industrial scale [40]. Similarly, after CNTs@pearl millet stem were employed for the remediation of Safranin O and MB, a mixture of hot water and acetone was used to regenerate the GCNM. The results showed that in the first round, the remediation efficiencies were 98.81% and 98.11%, and these marginally diminished to 94.45% and 93.45% in the tenth round for Safranin O and MB, respectively. This demonstrates once more how well green carbon nanoparticles regenerate and may be reused. The occupation of some GCNM active sites by dye pollutant molecules during the desorption process was attributed to the slight but gradual decrease in the percentage removal of dye pollutants. Based on these results, this GCNM may be utilized on an industrial scale without losing its remediation effectiveness [46].

5.8 Pilot-scale potential and application in real effluent

Notably, investigating the performance of GCNMs in real contaminated water is vital in establishing their potential for industrial-scale applications beyond the laboratory [22, 92, 107]. This section empirically discusses important research findings in this direction. For instance, one study explored the potential of green N-doped CDs@alginate bead hydrogel to remove Cu(II) from real wastewater obtained from Beijing city and Anhui province, China. In this study, 1 g l^{-1} of the GCNM was introduced into real wastewater containing 5 mg l^{-1} of Cu(II). The research findings showed that this material has the potential to reduce the copper concentration to the World Health Organization (WHO) and Environmental Protection Agency (EPA) acceptable limits of 0.8 mg l^{-1} for tap water and 1.05 mg l^{-1} for wastewater. It was opined that the different concentrations of Cu(II) in the two water matrices could be ascribed to interference from coexisting cations such as magnesium and calcium ions [108]. A small impact due to interference was also observed in the interaction between dye pollutants and the surface of peanut shell BC@GO in river and industrial wastewater in the remediation of bromophenol blue and CV. This is consistent with the report by Peng's research group [65]. Specifically, when sewage sludge CQDs@Bi$_2$MoO$_6$@Bi$_2$S$_3$ were employed as a catalyst for the removal of tetracycline (TC) from tap water, Xiangjiang River water, and Taozi Lake water, the cleanup efficiency of the GCNMs was not significantly affected, demonstrating that although other background ions found in real water systems could have competed with the GCNM, it still maintained good remediation efficacy [65]. The findings demonstrated that this GCNM has great prospects as a green and inexpensive filter material for effective wastewater treatment systems [37, 65].

In another study [109], *S. munja* biomass-functionalized CNTs@CTAB were investigated for their potential in real dye effluent treatment. In this study, real dye wastewater samples, which had been directly discharged into sewers, were obtained from two different sites at local market dyeing shops in the Bhiwani territory. The results showed that when 0.03 g of the GCNM was introduced into 10 ml of the concentrated effluent, a cleanup rate of over 95% was achieved for the two treated effluents, signifying the admirable potential of the GCNM for effective remediation of real-world effluent containing an assortment of pollutants [109]. In a comparable study [61], the applicability of pomegranate-peel-waste-derived NBC@MgAl-layered double hydroxide (LDH) for the remediation of MB from seawater, tap water, and industrial wastewater was investigated by spiking 20 ml of each of the three actual water matrices with 5–15 mg l^{-1} of MB dye pollutant, followed by the introduction of 15 mg GCNM at pH 7 for around 25 min. Interestingly, for the three spiked real water systems, remediation efficiencies of more than 92% were recorded [61], which is consistent with the results obtained when waste tobacco leaf CDs were used to detect and remove Hg^{2+} from lake, pond, and tap water [90]. Kaushik's research team [110] explored the use of pear fruit graphene aerogel for the elimination of CV, MB, and RhB from simulated real dye effluent. Notably, 96% of MB and 70% of CV were removed. Unfortunately, the GCNM had lower removal efficiency for RhB, probably because of its larger size, combined with the sorption of other smaller background ions/pollutants present in the industrial effluent [110].

Furthermore, in recent work, the practicality of rice-straw-derived CDs for the removal of Pb(II) pollutants from river, tap, and reverse osmosis (RO)-spiked water samples was examined. The recovery rates reported ranged from 96% to 101% for the spiked real water systems, indicating that this GCNM demonstrates potential for use as a fluorescent probe in lead ion detection/removal [111].

In another study [92], CQDs@poly(anthranilic acid-formaldehyde-phthalic acid) were used to remove U(VI) from real water matrices (tap water, seawater, and wastewater). Specifically, the microwave-assisted sorption approach showed that the GCNMs had great potential for removing U(VI) from real water matrices in a short time of about 1 min. More specifically, the findings showed that the removal rates of U(VI) from wastewater, seawater, and tap water were 97.3%, 96.0%, and 80.9%, respectively [92]. Similarly, Malathion, a pesticide water pollutant, was removed from river and dam water samples using olive pomace graphene QDs@Fe$_3$O$_4$. It was discovered that this GCNM achieved remediation efficiencies of 73% and 67% in 210 min for the dam and river water, respectively. Although these efficiencies were slightly lower than that obtained for distilled water, these results demonstrate the robustness of this GCNM under harsh conditions and point to its potential applicability in real industrial water treatment processes [75]. Overall, it can be inferred from this section that GCNMs have great potential for pilot-scale cleanup of both organic and inorganic pollutants.

5.9 Challenges and future research scope

Notably, this chapter has revealed that GCNMs are game changers in the realm of environmental remediation. However, some salient challenges that can serve as future research hotspots are identified and highlighted below.

Scalability, field trials, and pilot studies: there are only a few studies involving field trials of GCNMs for pollutant removal; thus, it is necessary to progress beyond the lab scale to real-world applications, such as wastewater plants, polluted rivers, and mine tailings.

Environmental and toxicological risk: although there are claims that GCNMs are eco-friendly, their fate, transport, and ecotoxicity in real environmental systems are not well understood. Therefore, thorough life cycle assessments and risk analyses are necessary to guarantee greater safety.

Regeneration and reusability techniques: the findings discussed in this review show that GCNMs have good reusability potential; nevertheless, future researchers should focus on green regeneration strategies such as solar heating and bioregeneration to boost sustainability.

Disposal of recycled spent GCNMs and used eluent: the final destination of GCNMs after they have been reused repeatedly is still unknown. Also, the disposal of eluents/regenerants used to restore the spent GCNMs is still a problematic issue. If the disposal of recycled spent GCNMs and used eluents is not properly handled, they could become an environmental nuisance and foil the original aim of win-win environmental remediation.

Techno-economic analysis: the cost of the entire remediation process from the synthesis of GCNMs to application, even if it is just at the lab scale, is often neglected, and this is an important area of interest for various stakeholders, including engineers. Thus, future research should direct more effort in this direction.

Hybrid technology: combining two or three remediation methods, as in the case of the photocatalytic membrane-based technique, can improve the synergetic remediation performance of GCNMs.

Integration with artificial intelligence (AI): there is a need for researchers to channel more energy into integration with AI, including machine learning, neural networks, and statistical physics modeling for improved process optimization and time savings.

Regulatory frameworks and risk assessment models: these involve partnering with policymakers and other environmental stakeholders to design and develop regulations for the safe use and disposal of GCNMs.

5.10 Conclusions

This chapter explored the utilization of GCNMs for the removal of both organic and inorganic pollutants. The GCNMs used were either CDs, graphene/GO, CNTs, or bio/hydrochar-based nanomaterials. By and large, various GCNMs showed good pollutant removal capacity of over 70% on average and reached 100% in some cases at pH values between 2 and 10. However, it was observed that functionalized or doped GCNMs had better remediation performance in the shortest possible time. Furthermore, the review findings showed that GCNMs can be reused for up to two to ten times with an average removal efficiency of 70%–100% while maintaining their structural integrity. This revealed that GCNMs are economically suitable for actual wastewater treatment and can be commercially employed in industrial applications. Other studies of the pilot-scale potential of GCNMs and their application in real effluent were underscored. Areas of future research were highlighted, including integration with neural networks/artificial intelligence, exploration of field trial scenarios, and disposal of spent GCNMs.

References

[1] Tahir M B, Nawaz T, Nabi G, Sagir M, Khan M I and Malik N 2022 Role of nanophotocatalysts for the treatment of hazardous organic and inorganic pollutants in wastewater *Int. J. Environ. Anal. Chem.* **102** 491–515

[2] Ethiraj S and Samuel M S 2024 A comprehensive review of the challenges and opportunities in microalgae-based wastewater treatment for eliminating organic, inorganic, and emerging pollutants *Biocatal. Agric. Biotechnol.* **60** 103316

[3] Tatarchuk T, Bououdina M, Al-Najar B and Bitra R B 2019 Green and ecofriendly materials for the remediation of inorganic and organic pollutants in water *A New Generation Material Graphene: Applications in Water Technology* ed M Naushad (Cham: Springer) 69–110

[4] Rivas B L, Urbano B F and Sánchez J 2018 Water-soluble and insoluble polymers, nanoparticles, nanocomposites and hybrids with ability to remove hazardous inorganic pollutants in water *Front. Chem.* **6** 320

[5] Emenike E C, Emmanuel S S, Iwuozor K O, Okwu K C and Adeniyi A G 2024 12 - Plant biomass materials in petrochemical application *Plant Biomass Applications* ed M Jawaid, A Khan and A M A Asiri (New York: Academic) 351–83

[6] Dubovina M, Krčmar D, Grba N, Watson M A, Rađenović D, Tomašević-Pilipović D and Dalmacija B 2018 Distribution and ecological risk assessment of organic and inorganic pollutants in the sediments of the transnational Begej canal (Serbia-Romania) *Environ. Pollut.* **236** 773–84

[7] Baluch M A and Hashmi H N 2019 Investigating the impact of anthropogenic and natural sources of pollution on quality of water in Upper Indus Basin (UIB) by using multivariate statistical analysis *J. Chem.* **2019** 4307251

[8] Kalsoom A, Ali S, Khan N, Ali M A and Khan 2024 Enhanced ultrasonic adsorption of pesticides onto the optimized surface area of activated carbon and biochar: adsorption isotherm, kinetics, and thermodynamics *Biomass Convers. Biorefinery* **14** 15519–34

[9] Panis C, Candiotto L Z P, Gaboardi S C, Gurzenda S, Cruz J, Castro M and Lemos B 2022 Widespread pesticide contamination of drinking water and impact on cancer risk in Brazil *Environ. Int.* **165** 107321

[10] Raabe H A, Costin G-E, Allen D G, Lowit A, Corvaro M, O'Dell L, Breeden-Alemi J, Page K, Perron M and Flint Silva T 2025 Human relevance of *in vivo* and *in vitro* skin irritation tests for hazard classification of pesticides *Cutan. Ocul. Toxicol.* **44** 1–21

[11] Emmanuel S S, Adesibikan A A, Bayode A A, Olawoyin C O, Isukuru E J and Raji O Y 2024 A review on covalent organic frameworks with Mult-site functional groups as superior adsorbents for adsorptive sequestration of radio-contaminants *J. Organomet. Chem.* **1015** 123226

[12] Gupta K, Aggarwal R, Sharma M, Yadav R, Gupta R, Westman G and Sonkar S K 2025 Microcrystalline cellulose-based, nitrogen-doped carbon nanoflakes for adsorption of uranium and thorium *React. Chem. Eng.* **10** 1767–75

[13] Hu Y, Xie Z, Zou C and Tang W 2025 Highly efficient adsorption of Pb (II) by Cucurbit [7] uril-modified amino-expanded graphite: experiments, synthetic modelling optimisation and DFT calculations *Sep. Purif. Technol.* **363** 132177

[14] Badamasi H, Sanni S O, Ore O T, Bayode A A, Koko D T, Akeremale O K and Emmanuel S S 2024 Eggshell waste materials-supported metal oxide nanocomposites for the efficient photocatalytic degradation of organic dyes in water and wastewater: a review *Bioresour. Technol. Rep.* **26** 101865

[15] Liu P, Lyu J and Bai P 2025 Synthesis of mixed matrix membrane utilizing robust defective MOF for size-selective adsorption of dyes *Sep. Purif. Technol.* **354** 128672

[16] Bayode A A, Emmanuel S S and Terlanga K D 2025 Chapter 19—nanomaterials and their derivative biocomposites for dye adsorption *Engineered Biocomposites Dye Adsorption* ed A H Jagaba, S R Mohamed Kutty, M H Isa and A H Birniwa (Amsterdam: Elsevier) 313–37

[17] Bayode A A, Sunday Emmanuel S, Sanni S O, Lakhdar F, Fu L, Shang J and Shawn Fan H-J 2025 Biogenic fabrication of spinel nickel ferrite imprinted on Bifurcaria bifurcata Macro-Alga activated carbon for the adsorption of ciprofloxacin and metronidazole *Chem. Eng. Sci.* **302** 120843

[18] Bayode A A, Emmanuel S S, Osti A, Olorunnisola C G, Egbedina A O, Koko D T, Adedipe D T, Helmreich B and Omorogie M O 2024 Applications of perovskite oxides for the cleanup and mechanism of action of emerging contaminants/steroid hormones in water *J. Water Process Eng.* **58** 104753

[19] Emmanuel S S, Aniekezie F C and Adesibikan A A 2024 Covalent organic frameworks and metal–organic frameworks for sustainable adsorptive removal/extraction of dirty dozen chemicals: a review *J. Chinese Chem. Soc.* **71** 978–1007

[20] Isukuru E J, Opha J O, Isaiah O W, Orovwighose B and Emmanuel S S 2024 Nigeria's water crisis: abundant water, polluted reality *Clean. Water* **2** 100026

[21] Emmanuel S S, Badamasi H, Sanni S O, Ore O T, Bayode A A and Adesibikan A A 2025 A comprehensive review on photocatalytic degradation of agro-organochlorine pollutants using multifunctional metal oxide and supported and doped metal oxide nanoarchitecture materials *J. Chinese Chem. Soc.* **72** 265–305

[22] Emmanuel S S, Adesibikan A A, Ore O T, Bayode A A, Badamasi H, Sanni S O and Ilo O P 2025 A comprehensive review on biomass waste-derived biochar for sustainable adsorptive remediation of hazardous radio-contaminants *Waste Biomass Valori.* **16** 2029–73

[23] Rani M and Shanker U 2022 Green nanomaterials: an overview *Green Functionalized Nanomaterials for Environmental Applications* (Amsterdam: Elsevier) 43–80

[24] Emmanuel S S, Adesibikan A A, Saliu O D and Opatola E A 2023 Greenly biosynthesized bimetallic nanoparticles for ecofriendly degradation of notorious dye pollutants: a review *Plant Nano Biol.* **3** 100024

[25] Iwuozor K O *et al* 2025 Repurposing spent sugarcane bagasse biosorbent from waste lubricating oil spill into biochar *Sugar Tech.* **27** 1290–9

[26] Emmanuel S S, Adesibikan A A, Olawoyin C O and Idris M O 2024 Photocatalytic degradation of maxilon dye pollutants using nano-architecture functional materials: a review *ChemistrySelect* **9** e202400316

[27] Sabet M and Mahdavi K 2019 Green synthesis of high photoluminescence nitrogen-doped carbon quantum dots from grass via a simple hydrothermal method for removing organic and inorganic water pollutions *Appl. Surf. Sci.* **463** 283–91

[28] Beker S A, Truskewycz A, Cole I and Ball A S 2020 Green synthesis of Opuntia-derived carbon nanodots for the catalytic decolourization of cationic dyes *New J. Chem.* **44** 20001–12

[29] Malavika J P, Shobana C, Ragupathi M, Kumar P, Lee Y S, Govarthanan M and Selvan R K 2021 A sustainable green synthesis of functionalized biocompatible carbon quantum dots from Aloe barbadensis Miller and its multifunctional applications *Environ. Res.* **200** 111414

[30] Ying T Y, Raman A A A, Bello M M and Buthiyappan A 2020 Magnetic graphene oxide-biomass activated carbon composite for dye removal *Korean J. Chem. Eng.* **37** 2179–91

[31] Barooah P, Mushahary N, Das B and Basumatary S 2024 Waste biomass-based graphene oxide decorated with ternary metal oxide (MnO–NiO–ZnO) composite for adsorption of methylene blue dye *Clean. Water* **2** 100049

[32] Prasannan A and Imae T 2013 One-pot synthesis of fluorescent carbon dots from orange waste peels *Ind. Eng. Chem. Res.* **52** 15673–8

[33] Mallikarjuna K, Vattikuti S V P, Manne R, Manjula G, Munirathnam K, Mallapur S, Marraiki N, Mohammed A, Reddy L V and Rajesh M 2021 Sono-chemical synthesis of silver quantum dots immobilized on exfoliated graphitic carbon nitride nanostructures using ginseng extract for photocatalytic hydrogen evolution, dye degradation, and antimicrobial studies *Nanomaterials* **11** 2918

[34] Goswami S, Banerjee P, Datta S, Mukhopadhayay A and Das P 2017 Graphene oxide nanoplatelets synthesized with carbonized agro-waste biomass as green precursor and its application for the treatment of dye-rich wastewater *Process Saf. Environ. Prot.* **106** 163–72

[35] Gan L, Li B, Chen Y, Yu B and Chen Z 2019 Green synthesis of reduced graphene oxide using bagasse and its application in dye removal: a waste-to-resource supply chain *Chemosphere* **219** 148–54

[36] Nugraha M W, Abidin N H Z and Sambudi N S 2021 Synthesis of tungsten oxide/amino-functionalized sugarcane bagasse derived-carbon quantum dots (WO_3/N-CQDs) composites for methylene blue removal *Chemosphere* **277** 130300

[37] Sohni S, Gul K, Shah J A, Iqbal A, Sayed M and Khan S B 2023 Immobilization performance of graphene oxide-based engineered biochar derived from peanut shell towards cationic and anionic dyes *Ind. Crops Prod.* **206** 117656

[38] Zhang W, Yang Q, Luo Q, Shi L and Meng S 2020 Laccase-carbon nanotube nanocomposites for enhancing dyes removal *J. Clean. Prod.* **242** 118425

[39] Ahamad Z, Mashkoor F, Nasar A and Jeong C 2025 Multi-walled carbon nanotubes/TiO₂/ Chitosan nanocomposite for efficient removal of malachite green dye from aqueous system: a comprehensive experimental and theoretical investigation *Int. J. Biol. Macromol.* **295** 139461

[40] Yadav A, Bagotia N, Yadav S, Sharma A K and Kumar S 2021 Adsorptive studies on the removal of dyes from single and binary systems using *Saccharum munja* plant-based novel functionalized CNT composites *Environ. Technol. Innov.* **24** 102015

[41] El-Newehy M H, Aldalbahi A, Thamer B M and Abdulhameed M M 2024 Green and eco-friendly scalable synthesis of chitosan-carbon nanocomposite for efficient dye removal *Diam. Relat. Mater.* **148** 111461

[42] Ai L and Li L 2013 Efficient removal of organic dyes from aqueous solution with ecofriendly biomass-derived carbon@montmorillonite nanocomposites by one-step hydro-thermal process *Chem. Eng. J.* **223** 688–95

[43] Ai L and Jiang J 2012 Removal of methylene blue from aqueous solution with self-assembled cylindrical graphene–carbon nanotube hybrid *Chem. Eng. J.* **192** 156–63

[44] Inyang M, Gao B, Zimmerman A, Zhang M and Chen H 2014 Synthesis, characterization, and dye sorption ability of carbon nanotube–biochar nanocomposites *Chem. Eng. J.* **236** 39–46

[45] Jain N, Basniwal R K, Suman , Srivastava A K and Jain V K 2010 Reusable nanomaterial and plant biomass composites for the removal of Methylene Blue from water *Environ. Technol.* **31** 755–60

[46] Yadav S, Yadav A, Bagotia N, Sharma A K and Kumar S 2023 Novel composites of Pennisetum glaucum with CNT: preparation, characterization and application for the removal of safranine O and methylene blue dyes from single and binary systems, Biomass Convers *Biorefinery* **13** 16925–42

[47] Jiang W, Zhang L, Guo X, Yang M, Lu Y, Wang Y, Zheng Y and Wei G 2021 Adsorption of cationic dye from water using an iron oxide/activated carbon magnetic composites prepared from sugarcane bagasse by microwave method *Environ. Technol.* **42** 337–50

[48] Yadav S K, Dhakate S R and Pratap Singh B 2022 Carbon nanotube incorporated eucalyptus derived activated carbon-based novel adsorbent for efficient removal of methylene blue and eosin yellow dyes *Bioresour. Technol.* **344** 126231

[49] Azin E, Moghimi H and Taheri R A 2017 Development of carbon nanotube-mycosorbent for effective Congo red removal: optimization, isotherm and kinetic studies *Desalin. Water Treat.* **94** 222–30

[50] Zhai G, Liu F, Xiang H, Hu Z, Yu S, Jiang Y and Zhu M 2024 Heteroatom-doped carbon nanotube membranes from waste biomass for photoelectro-catalytic water purification *Adv. Sustain. Syst.* **8** 2300509

[51] Maddu A, Meliafatmah R and Rustami E 2021 Enhancing photocatalytic degradation of methylene blue using ZnO/carbon dots nanocomposite derived from coffee grounds *Polish J. Environ. Stud.* **30** 273–82

[52] Sekar A and Yadav R 2021 Green fabrication of zinc oxide supported carbon dots for visible light-responsive photocatalytic decolourization of Malachite Green dye: optimiza-tion and kinetic studies *Optik (Stuttg)* **242** 167311

[53] Ahmadian-Fard-Fini S, Salavati-Niasari M and Safardoust-Hojaghan H 2017 Hydrothermal green synthesis and photocatalytic activity of magnetic CoFe 2 O 4–carbon quantum dots nanocomposite by turmeric precursor *J. Mater. Sci., Mater. Electron.* **28** 16205–14

[54] Zaib M, Arshad A, Khalid S and Shahzadi T 2023 One pot ultrasonic plant mediated green synthesis of carbon dots and their application in visible light induced dye photocatalytic studies: a kinetic approach *Int. J. Environ. Anal. Chem.* **103** 5063–81

[55] Kusumawati E, Agustin A R, Widiyanti E, Hayati A N and Lumintu D 2019 Study of carbon nanodots from water hyacinth (Eichornia crassipes) to degrade textiles dyes of Skycion yellow HE-4R *IOP Conf. Ser.: Mater. Sci. Eng.* 509 *(Bristol: IOP Publishing)* 12013

[56] Shivalkar S, Gautam P K, Verma A, Maurya K, Sk M P, Samanta S K and Sahoo A K 2021 Autonomous magnetic microbots for environmental remediation developed by organic waste-derived carbon dots *J. Environ. Manage.* **297** 113322

[57] Waseem Basha Z, Muniraj S and Senthil Kumar A 2024 Neem biomass-derived carbon quantum dots synthesized via one-step ultrasonification method for ecofriendly methylene blue dye removal *Sci. Rep.* **14** 9706

[58] Ali S M, Ashour B, Farahat M G and El-Sherif R M 2024 Biomass-based perovskite/ graphene oxide composite for the removal of organic pollutants from wastewater *Ceram. Int.* **50** 49085–94

[59] Chakraborty V, Das P and Roy P K 2023 Lanthanum oxide–graphene oxide coated functionalized pyrolyzed biomass from sawdust and its application for dye removal present in solution *Biomass Convers. Biorefinery* **13** 5601–10

[60] Chen H, Liu T, Meng Y, Cheng Y, Lu J and Wang H 2020 Novel graphene oxide/aminated lignin aerogels for enhanced adsorption of malachite green in wastewater *Colloids Surf. A: Physicochem. Eng. Asp.* **603** 125281

[61] Nabil G M and Mahmoud M E 2024 Superior adsorptive uptake of methylene blue pollutant by nanobiochar-impregnated-layered double hydroxides *Inorg. Chem. Commun.* **168** 112913

[62] Mohiuddin I, Singh R and Kaur V 2024 Blending polydopamine-derived imprinted polymers with rice straw-based fluorescent carbon dots for selective detection and adsorptive removal of ibuprofen *Int. J. Biol. Macromol.* **269** 131765

[63] Chen S, Zhang S-Z and Jiang H 2024 Modification of crystal-optimized TiO_2 with biomass-derived carbon quantum dots for highly efficient degradation of favipiravir in water *ACS ES&T Water* **4** 531–42

[64] Spaolonzi M P, Duarte E D V, Oliveira M G, Costa H P S, Ribeiro M C B, Silva T L, Silva M G C and Vieira M G A 2022 Green-functionalized carbon nanotubes as adsorbents for the removal of emerging contaminants from aqueous media *J. Clean. Prod.* **373** 133961

[65] Peng P, Chen Z, Li X, Wu Y, Xia Y, Duan A, Wang D and Yang Q 2022 Biomass-derived carbon quantum dots modified Bi_2MoO_6/Bi_2S_3 heterojunction for efficient photocatalytic removal of organic pollutants and Cr (VI) *Sep. Purif. Technol.* **291** 120901

[66] Yu Y, Zhu Z, Liu Z, Dong H, Liu Y, Wei M, Huo P, Li C and Yan Y 2019 Construction of the biomass carbon quantum dots modified heterojunction Bi_2WO_6/Cu_2O photo-catalysis for enhancing light utilization and mechanism insight *J. Taiwan Inst. Chem. Eng.* **102** 197–201

[67] Li X, Yu Y, Wang Y, Di Y, Liu J, Li D, Wang Y, Zhu Z, Liu H and Wei M 2023 Withered magnolia-derived BCDs onto 3D flower-like Bi_2WO_6 for efficient photocatalytic TC degradation and CO_2 reduction *J. Alloys Compd.* **965** 171520

[68] Bayode A A, Folorunso M T, Helmreich B and Omorogie M O 2023 Biomass-tuned reduced graphene oxide@Zn/Cu: benign materials for the cleanup of selected nonsteroidal anti-inflammatory drugs in water *ACS Omega* **8** 7956–67

[69] Costa H P S, Duarte E D V, da Silva F V, da Silva M G C and Vieira M G A 2024 Green synthesis of carbon nanotubes functionalized with iron nanoparticles and coffee husk biomass for efficient removal of losartan and diclofenac: adsorption kinetics and ANN modeling studies *Environ. Res.* **251** 118733

[70] Ramanayaka S, Tsang D C W, Hou D, Ok Y S and Vithanage M 2020 Green synthesis of graphitic nanobiochar for the removal of emerging contaminants in aqueous media *Sci. Total Environ.* **706** 135725

[71] Hak C H, Leong K H, Chin Y H, Saravanan P, Tan S T, Chong W C and Sim L C 2020 Water hyacinth-derived carbon quantum dots and gC_3N_4 composites for sunlight-driven photodegradation of 2,4-dichlorophenol *SN Appl. Sci.* **2** 1–14

[72] Si Q, Guo W, Wang H, Liu B, Zhao Q, Luo H and Ren N 2021 Bio-CQDs surface modification BiOCl for the BPA elimination and evaluation in visible light: the contribution of C-localized level *J. Colloid Interface Sci.* **602** 1–13

[73] Cao S, Zhou Y, Xi C, Tang T and Chen Z 2021 Enhanced adsorption of malathion and phoxim by a three-dimensional magnetic graphene oxide-functionalized citrus peel-derived bio-composite *Anal. Methods* **13** 2951–62

[74] Mustafa S, Bhatti H N, Maqbool M, Khan A, Alraih A M and Iqbal M 2024 Renewable functional composites of algal biomass with graphene oxide and Na-alginate for the adsorptive removal of 2,4-D herbicide *Sustain. Chem. Pharm.* **39** 101577

[75] Assafi A, Zarouki M A, Hejji L, Aoulad El Hadj Ali Y, Chraka A, Pérez-Villarejo L, Sánchez-Soto P J, Souhail B and Azzouz A 2025 Olive pomace-derived graphene quantum dots decorated with iron oxide nanoparticles for efficient malathion removal from environmental water: theoretical and experimental studies *Diam. Relat. Mater.* **155** 112255

[76] Parambil A M, Priyadarshini E, Goutam R, Tsai P-C, Huang P-C, Rajamani P, Lin Y-C and Ponnusamy V K 2024 Self-assembled mesoporous silica decorated with biogenic carbon dot nanospheres hybrid nanomaterial for efficient removal of aqueous Methoxy-DDT via a Short-Bed Adsorption column technique *Environ. Res.* **260** 119653

[77] Thaveemas P, Chuenchom L, Kaowphong S, Techasakul S, Saparpakorn P and Dechtrirat D 2021 Magnetic carbon nanofiber composite adsorbent through green in-situ conversion of bacterial cellulose for highly efficient removal of bisphenol A *Bioresour. Technol.* **333** 125184

[78] Adebowale K O and Egbedina A O 2022 Facile green synthesis of bio-carbon material from eggshells and its application for the removal of Bisphenol A and 2,4,6-trichlorophenol from water *Environ. Nanotechnol. Monit. Manag.* **17** 100622

[79] Duarah R and Karak N 2017 Facile and ultrafast green approach to synthesize biobased luminescent reduced carbon nanodot: an efficient photocatalyst *ACS Sustain. Chem. Eng.* **5** 9454–66

[80] Ariyanti D, Lesdantina D, Purbasari A and Astuti Y 2023 Synthesis of graphene-like material derived from biomass from agricultural waste and its application in Cu (II) removal *Korean J. Chem. Eng.* **40** 964–74

[81] Ahmad A 2024 An innovative step to fabricate biomass-derived reduced graphene oxide electrodes to boost energy efficiency with metal removal using an electrochemical approach *Biomass Convers. Biorefin.* **15** 5997–6012

[82] Lai K C, Lee L Y, Hiew B Y Z, Thangalazhy-Gopakumar S and Gan S 2020 Facile synthesis of xanthan biopolymer integrated 3D hierarchical graphene oxide/titanium dioxide composite for adsorptive lead removal in wastewater *Bioresour. Technol.* **309** 123296

[83] Yaqoob A A, Serrà A, Ibrahim M N M and Yaakop A S 2021 Self-assembled oil palm biomass-derived modified graphene oxide anode: an efficient medium for energy transportation and bioremediating Cd (II) via microbial fuel cells *Arab. J. Chem.* **14** 103121

[84] Zohrabi Y, Ghazi M E, Izadifard M, Valipour A and Ayyaru S 2024 Resource utilization of oak fruit peel as biomass waste for the synthesis of carbon with graphene oxide-like composition and its composite with $Mg_{1-x}Ca_xFe_2O_4$ for Cd (ii) removal from water: characterization, magnetic properties, and potential adso *Environ. Sci. Water Res. Technol.* **10** 1920–37

[85] Samuel M S, Subramaniyan V, Bhattacharya J, Chidambaram R, Qureshi T and Pradeep Singh N D 2018 Ultrasonic-assisted synthesis of graphene oxide—fungal hyphae: an efficient and reclaimable adsorbent for chromium (VI) removal from aqueous solution *Ultrason. Sonochem.* **48** 412–7

[86] Tayyab M, Anwar S, Shafiq F, Shafique U, Kaya C and Ashraf M 2025 Adsorption isotherms and removal of lead (II) and cadmium (II) from aqueous media using nano-biochar and rice husk *Int. J. Phytoremediation* **27** 244–59

[87] Elbehiry F, Darweesh M, Al-Anany F S, Khalifa A M, Almashad A A, El-Ramady H, El-Banna A, Rajput V D, Jatav H S and Elbasiouny H 2022 Using biochar and nanobiochar of water hyacinth and black tea waste in metals removal from aqueous solutions *Sustainability* **14** 10118

[88] Perumal S, Atchudan R, Thirukumaran P, Yoon D H, Lee Y R and Cheong I W 2022 Simultaneous removal of heavy metal ions using carbon dots-doped hydrogel particles *Chemosphere* **286** 131760

[89] Issa M A, Zentou H, Jabbar Z H, Abidin Z Z, Harun H, Halim N A A, Alkhabet M M and Pudza M Y 2022 Ecofriendly adsorption and sensitive detection of Hg (II) by biomass-derived nitrogen-doped carbon dots: process modelling using central composite design *Environ. Sci. Pollut. Res.* **29** 86859–72

[90] Yang H, Su X, Cai L, Sun Z, Lin Y, Yu J, Hao L and Liu C 2022 Glutathione assisting the waste tobacco leaf to synthesize versatile biomass-based carbon dots for simultaneous detection and efficient removal of mercury ions *J. Environ. Chem. Eng.* **10** 108718

[91] Chen T, Shi P, Zhang J, Li Y, Duan T, Dai L, Wang L, Yu X and Zhu W 2018 Natural polymer konjac glucomannan mediated assembly of graphene oxide as versatile sponges for water pollution control *Carbohydr. Polym.* **202** 425–33

[92] Mahmoud M E, Fekry N A and Abdelfattah A M 2020 Removal of uranium (VI) from water by the action of microwave-rapid green synthesized carbon quantum dots from starch-water system and supported onto polymeric matrix *J. Hazard. Mater.* **397** 122770

[93] Zhang S, Arkin K, Zheng Y, Ma J, Bei Y, Liu D and Shang Q 2022 Preparation of a composite material based on self-assembly of biomass carbon dots and sodium alginate hydrogel and its green, efficient and visual adsorption performance for Pb^{2+} *J. Environ. Chem. Eng.* **10** 106921

[94] Yahaya Pudza M, Zainal Abidin Z, Abdul Rashid S, Md Yasin F, Noor A S M and Issa M A 2020 Eco-friendly sustainable fluorescent carbon dots for the adsorption of heavy metal ions in aqueous environment *Nanomaterials* **10** 315

[95] Yaqoob A A, Serrà A, Bhawani S A, Ibrahim M N, Khan A, Alorfi H S, Asiri A M, Hussein M A, Khan I and Umar K 2022 Utilizing biomass-based graphene oxide–polyaniline–Ag electrodes in microbial fuel cells to boost energy generation and heavy metal removal *Polymers (Basel)* **14** 845

[96] Naeimi A, Amini M and Okati N 2022 Removal of heavy metals from wastewaters using an effective and natural bionanopolymer based on Schiff base chitosan/graphene oxide *Int. J. Environ. Sci. Technol.* **19** 1301–12

[97] Osman A I, Blewitt J, Abu-Dahrieh J K, Farrell C, Al-Muhtaseb A H, Harrison J and Rooney D W 2019 Production and characterisation of activated carbon and carbon nanotubes from potato peel waste and their application in heavy metal removal *Environ. Sci. Pollut. Res.* **26** 37228–41

[98] Nejatpour M, Ünsür A M, Yılmaz B, Gül M, Ozden B, Barisci S and Dükkancı M 2025 Enhanced photodegradation of perfluorocarboxylic acids (PFCAs) using carbon quantum dots (CQDs) doped TiO_2 photocatalysts: a comparative study between exfoliated graphite and mussel shell-derived CQDs *J. Environ. Chem. Eng.* **13** 115382

[99] Varshney N, Tariq M, Arshad F and Sk M P 2022 Biomass-derived carbon dots for efficient cleanup of oil spills *J. Water Process Eng.* **49** 103016

[100] Queiroz R N, de T, Neves F, da Silva M G C, Mastelaro V R, Vieira M G A and Prediger P 2022 Comparative efficiency of polycyclic aromatic hydrocarbon removal by novel graphene oxide composites prepared from conventional and green synthesis *J. Clean. Prod.* **361** 132244

[101] Bayode A A, Olisah C, Emmanuel S S, Adesina M O and Koko D T 2023 Sequestration of steroidal estrogen in aqueous samples using an adsorption mechanism: a systemic scientometric review *RSC Adv.* **13** 22675–97

[102] Umeh C T, Akinyele A B, Okoye N H, Emmanuel S S, Iwuozor K O, Oyekunle I P, Ocheje J O and Ighalo J O 2023 Recent approach in the application of nanoadsorbents for malachite green (MG) dye uptake from contaminated water: a critical review *Environ. Nanotechnol. Monit. Manag.* **20** 100891

[103] Emmanuel S S, Olawoyin C O, Adesibikan A A and Opatola E A 2023 A pragmatic review on bio-polymerized metallic nano-architecture for photocatalytic degradation of recalcitrant dye pollutants *J. Polym. Environ.* **32** 1–30

[104] Shang Y, Kan Y and Xu X 2023 Stability and regeneration of metal catalytic sites with different sizes in Fenton-like system *Chinese Chem. Lett.* **34** 108278

[105] Sun J, Feng S and Feng S 2020 Hydrothermally synthesis of MWCNT/N-TiO_2/UiO-66-NH_2 ternary composite with enhanced photocatalytic performance for ketoprofen *Inorg. Chem. Commun.* **111** 107669

[106] Zhang P, O'Connor D, Wang Y, Jiang L, Xia T, Wang L, Tsang D C W, Ok Y S and Hou D 2020 A green biochar/iron oxide composite for methylene blue removal *J. Hazard. Mater.* **384** 121286

[107] Emmanuel S S and Adesibikan A A 2024 A review on photocatalytic degradation of aromatic organoarsenic compounds in aqueous environment using nanomaterials *J. Chinese Chem. Soc.* **71** 1130–53

[108] Wei T, Ni H, Ren X, Zhou W, Gao H and Hu S 2024 Fabrication of nitrogen-doped carbon dots biomass composite hydrogel for adsorption of Cu (II) in wastewater or soil and DFT simulation for adsorption mechanism *Chemosphere* **361** 142432

[109] Yadav A, Bagotia N, Yadav S, Sharma N, Sharma A K and Kumar S 2022 Environmental application of *Saccharum munja* biomass-derived hybrid composite for the simultaneous removal of cationic and anionic dyes and remediation of dye polluted water: a step towards pilot-scale studies *Colloids Surf. A Physicochem. Eng. Asp.* **650** 129539

[110] Kaushik J, Kumar V, Garg A K, Dubey P, Tripathi K M and Sonkar S K 2021 Bio-mass derived functionalized graphene aerogel: a sustainable approach for the removal of multiple organic dyes and their mixtures *New J. Chem.* **45** 9073–83

[111] Singh J, Bhattu M, Verma M, Bechelany M, Brar S K and Jadeja R 2025 Sustainable valorization of rice straw into biochar and carbon dots using a novel one-pot approach for dual applications in detection and removal of lead ions *Nanomaterials* **15** 66

IOP Publishing

Sustainable Carbon Nanomaterials and their Applications

Rafik Naccache and Adedapo O. Adeola

Chapter 6

Biomedical applications of carbon nanomaterials

Fernanda Maria Policarpo Tonelli, Christopher Santos Silva, Vinicius Marx Silva Delgado, Vitória de Oliveira Lourenço, Geicielly da Costa Pinto, João Vitor Nunes and Flávia Cristina Policarpo Tonelli

Due to its unique properties, carbon is a versatile element of special relevance to the nanotechnology field. Among the most relevant applications of carbon nanomaterials, such as carbon nanotubes (CNTs), nanographene and its oxide, and carbon nanodots (CDs), are those associated with biomedicine. These biomedical applications include the use of nanoscale structures not only for efficient diagnosis but also in therapeutic protocols, where the structures can serve as carriers in drug delivery or even act as active substances themselves. Carbon nanomaterials can be designed to optimize aspects associated with these uses, such as target specificity, efficiency, and reduced undesirable cytotoxicity. This chapter is dedicated to exploring the potential of these nanomaterials in biomedical applications and the future prospects for the field.

6.1 Introduction

Carbon (represented by the chemical symbol C) is an abundant and versatile element on Earth, capable of exhibiting different forms such as diamond, graphite, and coal (Notarianni *et al* 2016). This element can display sp, sp^2, or sp^3 hybridization, which gives it unique chemical and physical properties of human interest, resulting in important contributions to different fields of expertise, such as technology, particularly nanotechnology (Maiti *et al* 2019).

Carbon nanomaterials can be used, for example, to promote the nanoremediation of environmental pollution. The growing development of different types of industries that have not considered sustainability has led, and continues to lead, to extensive environmental contamination, resulting in an imbalance in the ecosystem that threatens the survival of living beings (Silva *et al* 2024). Pollutants can be

doi:10.1088/978-0-7503-6325-9ch6
6-1

Figure 6.1. Examples of important applications of carbon nanomaterials in the biomedical field.

inorganic (Tonelli *et al* 2020), organic molecules (Tonelli *et al* 2024), or pathogenic organisms (Lourenço *et al* 2024) that can be released into the air, water, and/or soil. Materials such as graphene and graphene oxide (GO), for example, can be applied to promote the efficient management of these contaminants and also assist in the recovery of nutrient-deficient soils (Ferreira *et al* 2022).

However, this chapter is dedicated to reviewing the biomedical applications of carbon nanomaterials, as these nanoscale materials are extremely useful in fields involving implants, bioimaging/biosensing, drug delivery, tissue engineering, diagnosis, wound healing, cancer treatment, and applications associated with antimicrobial activity, for example (Malode *et al* 2024) (figure 6.1). Graphene and its derivatives, CNTs, and fullerenes will receive special attention.

6.2 Biomedical applications of graphene, graphene oxide, and graphene quantum dots

Graphene is a carbon nanomaterial that deserves to be highlighted due to its applications in biomedicine (Tonelli *et al* 2015). It is a two-dimensional structure first produced by Nobel Prize winners Novoselov and Geim in 2004 (Novoselov *et al* 2004). The carbon atoms of this material, which exhibit sp^2 hybridization, are organized in hexagons (similar to a honeycomb), giving the material the form of a flexible and stable sheet (figure 6.2) (Riley and Narayan 2021).

The oxidation of graphite can allow for the production of other carbon nanomaterials such as GO (Jiříčková *et al* 2022), another monolayer containing chemical groups that include oxygen, such as carboxyls, epoxies, and hydroxyls; consequently, it can also contain carbon atoms with sp^3 hybridization (Nováček *et al* 2017). Graphene quantum dots (GQDs), zero-dimensional nanostructures, are, however, smaller carbon nanomaterials than those previously presented in this section. Their size commonly varies from 2 to 50 nm, and they exhibit fluorescence (Bacon *et al* 2014). Upon excitation, they can emit light with wavelengths ranging

Figure 6.2. Graphene. In the upper part of the figure, the structure of the nanomaterial is similar to that of a honeycomb (shown below).

from the visible region to the infrared region, depending on the nanomaterial's quantum confinement properties (Minsu *et al* 2020). This characteristic favors their use, especially in bioimaging and biosensing, for example.

Graphene does not exhibit good solubility in biological systems; this nanomaterial tends to aggregate and exhibit cytotoxicity. As a result, it needs to undergo functionalization before being used in drug delivery (Wang *et al* 2011). The variety of possibilities for modifying graphene's surface, combined with the properties of this material, presents an opportunity to achieve stimuli-responsive drug delivery, triggered by pH, sound, temperature, or a magnetic field, for example (Khakpour *et al* 2023). As a carrier material, GO, which exhibits different chemical groups on its surface, is commonly preferred over graphene, as it also offers better solubility (Yang *et al* 2015). GQDs have been more recently applied as carriers compared to graphene and its oxide. Similar to GO, they also display functional groups on their surfaces that favor the task of delivering drugs. The biocompatibility and small dimensions of this nanomaterial are interesting features, making it capable of passing the blood–brain barrier (Henna and Pramod 2020).

As recent examples of these nanostructures delivering drugs, it is possible to highlight pristine few-layer graphene delivering diclofenac in a thermo-responsive way as the temperature was increased from 25 °C to 44 °C, allowing the hydrogel scaffold containing the nanomaterial to release the drug (Mauri *et al* 2021). A co-polymeric hydrogel containing graphene, chitosan, and polyvinyl alcohol could also deliver more than 97% of the drug methotrexate in 6 h. It proved to be non-hemolytic and blocked HepG2 growth *in vitro* (Mansha *et al* 2024). Doxorubicin, which exhibits an anticancer effect, was loaded more efficiently in functionalized graphene than in unmodified graphene (Zainal-Abidin *et al* 2020).

GQDs were used to modify polyethyleneimine to create a positively charged structure to deliver doxorubicin, aiming to negatively affect HCT116 cancer cells.

This delivery system was combined with another containing pyrenebutyric acid and hyaluronic acid to deliver the hydrophobic drug TAK-632, targeting the same cells. This dual delivery system was efficient not only *in vitro* but also *in vivo* (in mice) (Lin *et al* 2024). Recently, GQDs have also proven efficient in solving the problem of loading exosomes with drugs to be delivered to cancer cells. A drug loading efficiency of 66.3%, which surpassed the efficiency offered by other techniques, was exhibited for the delivery of doxorubicin (Zhang *et al* 2023). As this type of quantum dot exhibits antioxidant and anti-inflammatory activities, it could be used in a drug delivery system aimed at treating age-related macular degeneration with choroidal neovascularization. The nanomaterial was associated with an octadecyl-modified peptide (as a target to be cleaved by matrix metalloproteinase 9, commonly present during disease development) to deliver minocycline to a mouse model. The system's efficiency was optimized when it was associated with bevacizumab (Huang *et al* 2023).

GO modified with polyethylene glycol (PEG) was associated with gold nanoparticles to provide doxorubicin delivery that was responsive to pH and capable of offering the expected anticancer activity (Samadian *et al* 2020). Nanocolloids of this nanomaterial could be conjugated to the protein albumin from bovine serum to deliver Dabrafenib and Trichostatin A to melanoma cells, optimizing the effects on the target and attenuating the undesirable cytotoxicity naturally exhibited by the nanomaterial toward normal cells (Sima *et al* 2020). The reduced version of GO was added to a chitosan hydrogel to provide a carrier responsive to near-infrared light. The system was able to deliver teriparatide, which supports bone regeneration in rats (Wang *et al* 2021). Nanofibers of polyvinyl alcohol containing GO modified with silver nanoparticles delivered curcumin to assist in wound healing and prevent infection by bacteria, favoring cell proliferation and regeneration (Rahmani *et al* 2022). Chitosan beads containing reduced GO (rGO) offered a pH-responsive vehicle that delivered 5-fluorouracil and curcumin to fight MCF-7 cancer cells efficiently (Boddu *et al* 2022). This type of graphene could also be used to generate a membrane to deliver fucoxanthin. To provide a reactive oxygen species (ROS)-responsive delivery (triggered by H_2O_2), the carrier was associated with the copolymer PEG diacrylate 1,2-ethanedithiol (Wu *et al* 2022). Aspirin and doxorubicin were delivered to MDA-MB 231 cells (found in cases of breast cancer) *in vitro*, using a system made from GO nanoparticles (Sahoo *et al* 2022, Youn *et al* 2017, Mohajeri *et al* 2019). A hydrogel containing agarose, chitosan, and graphene was produced to deliver 5-fluorouracil with the intention of fighting breast cancer. The system was pH-sensitive and performed a sustained release at pH 5.4, releasing a large amount of the drug within 48 h; MCF-7 cancer cells then presented a significant decrease in viability (Rajaei *et al* 2023). Caffeic acid was delivered by β-cyclodextrin associated with GO and conjugated to Fe-based metal–organic frameworks; at an acidic pH of 5, a release rate of more than 57% was observed, negatively affecting A549 lung cancer cells. A dose of less than 400 $\mu g\ ml^{-1}$ affects normal HEK293 cells (Sontakke *et al* 2024).

GQDs, due to the fact that they can exhibit photoluminescence, are commonly applied in sensors to detect various substances. When immobilized on supports, the challenge of undesirable aggregation can be overcome efficiently. This nanomaterial was immobilized in a metal–organic framework containing zirconium to detect copper, a toxic heavy metal that represents a serious threat to human health (Y Chen *et al* 2024).

However, rGO can also be modified to allow detection. NO_2, for example, was efficiently detected with fast sensor recovery after a modification involving SnO_2 (Neri *et al* 2013). Alcohols were detected by a sensor containing thin layers of GO, based on the electrical response offered by the device (Moura *et al* 2023a).

The COVID-19 pandemic has motivated the production of a graphene-based biosensor to detect SARS-CoV-2. The transistor-based biosensor was produced using the nanomaterial combined with an antibody against the virus's spike protein. Electrical current modifications were noticed after exposure to human nasopharyngeal swabs containing the virus (Seo *et al* 2020). In addition, the use of graphene and GO in fabric/textile composites to offer antimicrobial properties (not only against viruses but also against bacteria) is receiving attention from scientists (Hu *et al* 2019, Noor *et al* 2019, Kumar *et al* 2020).

Graphene has already been successfully used in sensors to detect the muscle relaxant tizanidine hydrochloride. A nanocomposite containing the nanomaterial and zinc oxide nanorods (Jain *et al* 2016) and another containing graphene and silicon dioxide (Sinha *et al* 2015) performed the detection efficiently. Graphene-based sensors have already been studied for human health monitoring of asthma, chronic obstructive pulmonary disease, chronic kidney diseases, diabetes, gastric cancer, lung cancer, and sleep apnea, for example (Moura *et al* 2023b).

Graphene and its derivatives have already provided fluorescent emitters and quenchers to detect biomolecules and ions, for example (Zhu *et al* 2015). A stable composite containing GO functionalized by curcumin, exhibiting low cytotoxicity, allowed *in vivo* tumor biological imaging through photoluminescence (Xu *et al* 2017). It is interesting to highlight that chemical modification of GO's surface using PEG, gold nanoparticles, Fe_3O_4 nanoparticles, and folic acid optimized photoluminescence *in vitro* in three different fluorescence regions, which is an advantage in bioimaging (Esmaeili *et al* 2020). GO quantum dots exhibiting optical properties useful in bioimaging were produced using cost-effective raw materials, namely coal and ethanol, instead of CNTs, for example (Kang *et al* 2019).

Tissue engineering is another important field in which graphene and graphene-based nanomaterials are useful. They can be used to build scaffolds that mimic cells' ideal environment for proliferation and/or differentiation. Such scaffolds exhibit biocompatibility and stability, which promote efficient tissue regeneration (Resende *et al* 2014). Three-dimensional scaffolds containing GO and/or rGO, for example, supported stem cell differentiation to tissues of interest, such as cardiac tissue (in this case, the electrical conductivity of the nanomaterial is also advantageous) and bone. These nanomaterials are commonly applied in protein scaffolds to improve their desirable mechanical properties (Biru *et al* 2022). Many other properties important in biomedicine have already been demonstrated in graphene-based nanomaterials. Antiviral properties, for example, were detected in a GO biocompatible composite containing curcumin, which inhibited the infection caused by respiratory syncytial virus (Yang *et al* 2017). Thus, significant potential can already be explored with the aim of applying these materials in the biomedical field.

6.3 Biomedical applications of carbon nanotubes

CNTs are nanostructures generally formed from sp^2-hybridized carbon in a hexagonal formation, generating a tube-like structure (Murjani *et al* 2022). They can have one wall or multiple concentric walls (figure 6.3). CNTs were identified by Iijima in 1991 while studying an accumulation produced by the reaction between two carbon electrodes in a helium atmosphere. Soon after the discovery of CNTs, several theories and calculations were discussed to predict the different properties of this nanomaterial (Ebbesen 1994).

Nowadays, the properties of CNTs are well known, and they are characterized by high electrical conductivity, outstanding mechanical capabilities, thermal stability, unusual optical properties, and good chemical inertness (Norizan *et al* 2020). These properties make them great candidates for applications in the biomedical field, capable of being applied in various areas such as drug delivery, the diagnosis and monitoring of diseases, the treatment of diseases (including degenerative ones and cancer), and tissue engineering, for example.

Despite the unique properties of CNTs, their applicability *in vivo* is complex since they have poor solubility and dispersion in many solvents, making it necessary to improve not only their solubility but also biocompatibility by attaching different kinds of materials (surfactants, biomolecules, amines, and amino acids) to their structure; the process involved in this chemical modification of the surface is known as functionalization. Consequently, the functionalization of CNTs improves their

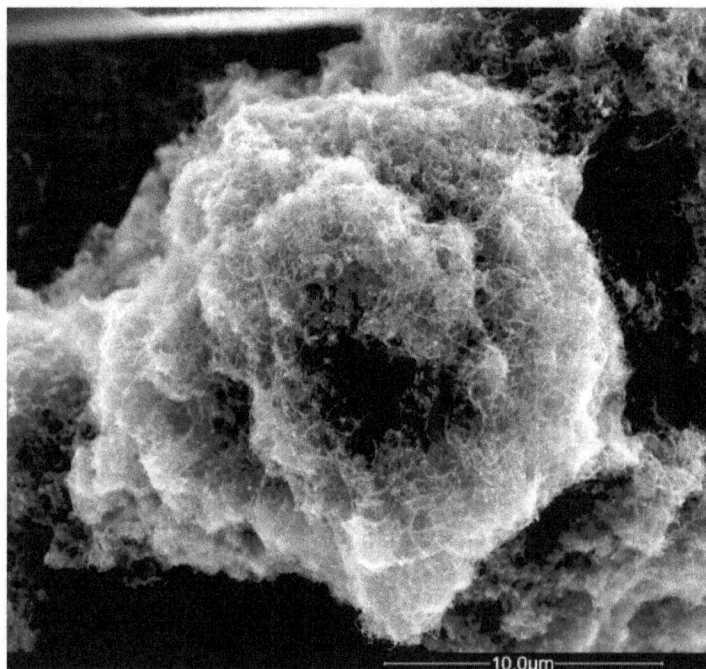

Figure 6.3. Field emission scanning electron microscopy image of a multiwalled CNT sample.

solubility and dispersion capacity, simultaneously enhancing the binding between various drugs or target cells and the nanotubes. (Karimi *et al* 2015). This optimizes and enables a large array of applications for these nanomaterials in the biomedical field (Murjani *et al* 2022).

Density functional theory (DFT) and time-dependent DFT (TDDFT) calculations showed that silicon-doped CNTs demonstrated improved properties for the delivery of remdesivir, a drug utilized for treating COVID-19. In this study, the nanomaterial was doped with silicon, which resulted in better conductivity, thermodynamics, and adsorption properties, leading, consequently, to better reactivity and interaction with remdesivir when compared with other doped and functionalized CNTs (Novir and Aram 2021). Neural cell death caused by the addictive use of methamphetamine was remedied by utilizing single-walled carbon nanotubes (SWCNTs) in conjunction with atorvastatin, a drug mainly used to lower blood cholesterol levels, which also has properties that induce cell growth and inhibit cell death. The SWCNTs were able to provide better transport of the conjugated medicine to the target cells, reducing the amount of atorvastatin needed for the treatment (Nikeafshar *et al* 2022). SWCNTs were also applied as a new drug delivery method when grafted with mesalazine and fluvoxamine, drugs used to treat inflammatory bowel diseases and mental disorders, respectively (Heidarian *et al* 2021).

Regarding drug delivery capacity, the use of CNTs in oncology treatments is one of the most extensively researched biomedical applications of these nanomaterials. PEG-grafted-[furfuryl-grafted-poly(styrene-alt-maleic anhydride)] functionalized SWCNTs exhibited a high loading capacity for the chemotherapy drug doxorubicin. The nanostructure effectively delivered the drug in a HeLa cancer cell environment, causing apoptosis of these cells at a physiological pH. Despite their cytotoxicity toward the HeLa cell line, they proved biocompatible when assayed in an environment containing normal human cells (Cao *et al* 2020). CNTs have also been able to improve the cytotoxic efficacy of doxorubicin when conjugated with a nanocomposite consisting of silica nanoparticles and DNA hydrogel. The drug loaded into this nanocomposite presented twice the efficiency of a nanocomposite consisting of only the nanoparticles and the hydrogel. This improved efficiency observed might be due to the way CNTs can easily cross cell membranes, allowing the anticancer drug to be internalized more efficiently by the cancer cells (Hu and Niemeyer 2020). Green-polyampholyte-functionalized SWCNTs also effectively loaded and released doxorubicin. The drug release profile was related to pH changes in the medium, with a significant release of the drug at a pH of 5.5, remaining stable at a pH of 7.4. In addition to drug release driven by pH variation, higher concentrations of the loaded nanostructure improved the cytotoxic effect on the HeLa cancer cell line, reducing the cancer cells' viability to 20.87% at 15 μg ml^{-1} (Thang 2020). Pemetrexed and quercetin in combination with multiwalled carbon nanotubes (MWCNTs) exhibited antitumoral potential against pancreatic cancer cells (PANC-1). The release of both substances by the MWCNTs decreased pancreatic cancer cell viability and induced a higher generation of ROS when compared to MWCNTs containing only one of the drugs. Furthermore, the study also demonstrated that codelivery of the substances resulted in better stability, internalization capacity, and oxidant activity against the PANC-1 cell line due to combinatorial effects between the two drugs (Badea *et al* 2020).

Theoretical studies involving computational quantum mechanics and calculations were also carried out to elucidate the behavior and properties of CNTs when bound to different chemotherapeutic drugs. To study the adsorption properties of cyto-phosphane on SWCNTs, Felegari and Hamedani applied DFT calculations to determine the electronic state of the structure. According to the results, the SWCNTs were sensitive to cytophosphane and exhibited effective drug adsorption, making them an acceptable candidate for applications in drug delivery (Felegari and Hamedani 2022). Ajeel *et al* used DFT calculations to investigate the drug delivery potential of SWCNTs doped with palladium to carry gemcitabine. This function-alization of SWCNTs enhanced the interaction with the drug, making the complex more stable and less reactive; it also contributed to an enhancement in the anticancer activity of gemcitabine (Ajeel *et al* 2023). Not only DFT but also the Hartree–Fock (HF) method has been applied to compare the interactions of busulfan, mercapto-purine, and fluorouracil with CNTs. Among the drugs, busulfan exhibited better structural stability when conjugated with the nanomaterial, as it had the most negative energy value. It also exhibited better solubility in an aqueous environment due to its higher dipole moment (Karachi 2023). However, the possibilities of using CNTs in the biomedical field are not limited to drug delivery.

The detection of biomolecules with traditional methodologies for early disease diagnostics and treatments is often seen as invasive, laborious, sluggish, and expensive. To overcome these disadvantages, the field of biosensors has vastly improved in the last decade, particularly with the development of biosensors based on nanomaterials. Among the nanomaterials, those based on carbon (such as CNTs) have attracted interest for the detection of biomolecules due to their singular physical, chemical, and electrical characteristics, especially sp^2 hybridization, which provides a vast surface-to-volume ratio, allowing better and faster sensitivity to biomolecules (Yang *et al* 2015, Lin *et al* 2019, Murjani *et al* 2022).

To monitor uric acid at the point-of-care level, a biosensor based on 3D super-aligned CNTs immobilized with the uricase enzyme was developed by Yang and coworkers. The biosensor exhibited high stability and sensitivity during the tests, allowing the researchers to continuously monitor the dynamic uric acid flow. Furthermore, when compared with an FDA-approved electrochemical analyzer using a paired t-test, the newly developed system proved to be reliable for detecting uric acid in real samples (Yang *et al* 2021). Horseradish peroxidase and glucose oxidase immobilized on an ionic liquid-functionalized graphene/CNT composite were used to develop a glucose biosensor. The graphene present in the structure contributed to the immobilization of the enzymes and provided a substantial contact area with the electrolyte. The CNTs improved the electron transfer process by diminishing the resistance from 1480 Ω to 550 Ω. Thus, the combination of these nanomaterials established a stable three-dimensional structure that exhibited good sensitivity and electrocatalytic response toward glucose (Zou *et al* 2020). To detect DNA with biomolecules, Li *et al* reported, for the first time, a peptide-based CNT thin-film transistor (TFT) biosensor. The peptide interacted with small pieces of nucleic acids, decreasing the zeta potential. The novel biosensor exhibited a better sensing range and detection limit when compared with commonly used DNA detectors and the NanoDrop (Thermo Scientific™) device. Additionally, cDNA from a breast cancer cell line (T47D) could also be detected and quantified by this peptide/CNT/TFT system (Li *et al* 2020). Han *et al* also developed a DNA biosensor through the use of signal amplification nanoclusters composed of CNTs and gold nanoparticles, presenting an 'urchinlike' tridimensional structure. Due to the intrinsic properties of CNTs and these nanoparticles, the sensor displayed a significant contact surface area and electronic conductivity, resulting in ultra-sensitive DNA detection and selectivity capabilities, selectively detecting mis-matched DNA sequences and noncomplementary sequences. Good stability and regeneration of the structure were also observed, and it maintained its performance over five cycles of use with minimal signal loss (Han *et al* 2020).

Dopamine released from C6 living cells could also be monitored in real time by Shu and coworkers using a stretchable electrochemical biosensor developed with a nickel metal–organic framework composite/gold nanoparticle-coated CNTs depos-ited on poly(dimethylsiloxane). High selectivity for dopamine detection was observed in the biosensor, as other endogenous substances, such as uric acid, salicylic acid, and glucose, did not significantly interfere with the results. A wide linear detection range of 50 nM to 15 μM and a high sensitivity of 1250 mA (cm^2 M)$^{-1}$ were also observed in the stretchable sensing platform, allowing real-time monitoring of the neurotransmitter in both stretched and unstretched states (Shu *et al* 2020). Phelane *et al* proposed polysulfone (PSF)-modified MWCNTs function-alized in nitric acid as a development base for detecting tyrosine. The MWCNT/PSF was deposited on a glassy carbon electrode (GCE) along with the tyrosinase enzyme (TyOx) to allow for the detection of tyrosine through its oxidation. The TyOx/ MWCNT/PSF/GCE biosensor exhibited low limits of detection and high sensitivity toward tyrosine when compared with similar biosensors reported previously in the literature (Phelane *et al* 2020). Huang and coworkers reported the detection of nicotinamide adenine dinucleotide (NADH) using an H_2O_2 biosensor developed with CNTs, tetrathiafulvalene, and horseradish peroxidase. The biosensor displayed good anti-interference capabilities, as other electroactive species like uric acid and glucose could not interfere with the sensor's detection range, sensitivity, and selectivity. Moreover, the response current for NADH determination increased proportionally to the NADH concentration in the medium, ranging from 10 to 790 μM (Huang *et al* 2020). Gulati *et al* used an MWCNT electrode functionalized with silane molecules to develop a resistive biosensor to detect P-glycoprotein, a biomarker of chronic myeloid leukemia. The functionalization of MWCNTs with silane provided binding sites for antibody biomolecules through covalent interac-tions, allowing the formation of a complex between the glycoprotein and the antibodies. The sensor proved to have desirable detection limits and was more cost-effective and faster than other sensors used to investigate chronic myeloid leukemia (Gulati *et al* 2020).

In bioimaging, CNTs are also useful (Acharya *et al* 2024). SWCNTs, for example, present a strong resonance in Raman scattering, which is an interesting aspect to explore. Liu and coworkers achieved success in the multicolor Raman imaging of living cells using an isotopically modified version of the nanomaterial (Liu *et al* 2008). Imaging of brown fat was also achieved using near-infrared photoluminescent SWCNTs coated with poly(2-methacryloyloxyethyl phosphoryl-choline-co-n-butyl methacrylate); the information obtained included thermogenic capability and capillary density (Yudasaka *et al* 2017). SWCNTs coated with a gel of PEG-based cross-linked polymers exhibited enhanced biocompatibility and bright fluorescence; 8.4 µg/mouse of this material was used in the high-quality imaging of vasculature (Nagai *et al* 2024). Multiwalled versions of CNTs have also proven useful in bioimaging; the liver and heart, for example, were scanned using MWCNTs as contrast agents (Delogu *et al* 2012).

Tissue engineering can also benefit from CNTs (Tonelli *et al* 2012), mitigating the risks associated with the transplant of biological material from a donor (Shi *et al* 2010) and offering optimized elastic moduli and tensile strength (Bosi *et al* 2014). A scaffold containing CNTs functionalized with polyglycerol sulfate coated onto electrospun polycaprolactone allowed the adhesion and proliferation of induced pluripotent stem cells, efficiently guiding neural differentiation (Xia *et al* 2019). More recently, sulfonated SWCNTs and 2,2,6,6-tetramethylpiperidinyl-1-oxyl (TEMPO)-oxidized cellulose nanofibrils generated a hydrogel that favored the expression of Connexin 43 and cardiac troponin-T proteins, an interesting option to consider when treating patients suffering from myocardial infarction (Sun *et al* 2023). Carboxylated MWCNTs in poly(caprolactone)/hydroxyapatite reinforced the conductivity of this scaffold and allowed MG-63 cell adhesion and survival, making this an interesting option for bone tissue engineering (Mohammadpour *et al* 2024).

6.4 Biomedical applications of fullerenes

Fullerenes (figure 6.4) are sphere-like structures made of sp^2-hybridized carbon atoms connected to three other neighboring atoms, creating a hollow sphere with only pentagonal and hexagonal faces. The most famous fullerene molecule, the C_{60} Buckminster fullerene, is named due to its similarity to the work of the architect

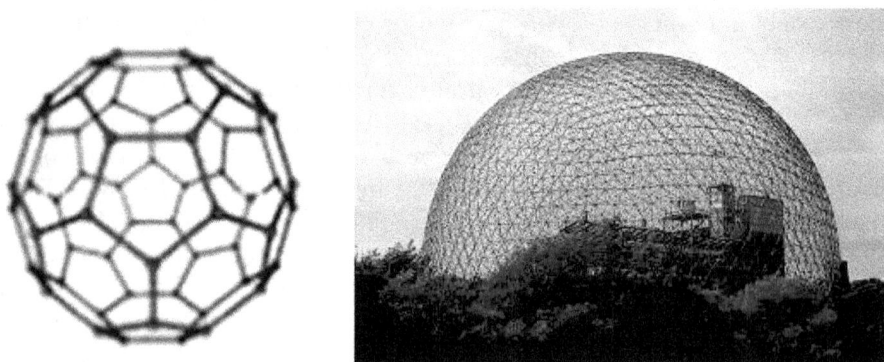

Figure 6.4. Fullerenes: (left) a fullerene molecule and (right) Buckminster's geodesic dome. Reproduced from [58]. CC BY-SA 2.0.

Buckminster Fuller, who built 'geodesic domes' (figure 6.4). Due to their shape, hydrophobicity, stability, chemical properties, and electronic configurations, fullerenes are seen as promising materials in nanomedicine (Barhoum *et al* 2019, Gaur *et al* 2021).

Due to their innate hydrophobicity, fullerenes are not soluble in water or other polar solvents but are soluble in halogenated and alkyl-substituted benzene, CS_2, 1,2-dichlorobenzene, and naphthalene. However, their surfaces can be functionalized in many different ways to change their characteristics and biological activities. For example, their solubility in water can be increased by attaching hydroxyl groups to their surface (Tzirakis and Orfanopoulos 2013, Bogdanović and Djordjević 2016), as illustrated in figure 6.5. So far, both the underivatized fullerenes and their derived counterparts have found interesting applications in the biomedical field; in drug delivery, for example.

Researchers have already reported the interactions between fullerenes and biological molecules and explored their benefits in biomedicine. Friedman *et al* and Nakamura and coworkers have studied the interaction between fullerene derivatives and HIV's protease. Friedman built a fullerene *in silico* using the DOCK3 program and described its interactions with the enzyme active site (Friedman *et al* 1993). Nakamura synthesized two fullerene derivatives with one or two acetoxyhydroxyl polar groups that displayed more affinity for the protein than their native fullerene counterpart; hence, they showed that the functional groups on the fullerene's surface interfere directly with the particle's interactions and, consequently, its activity (Nakamura *et al* 1996, Wang *et al* 2019).

However, it is wise to approach this subject with caution, as these carbon allotropes are toxic to living beings. Madannejad and coworkers reviewed the toxicity of carbon nanomaterials. They pointed out that one of the main aspects of

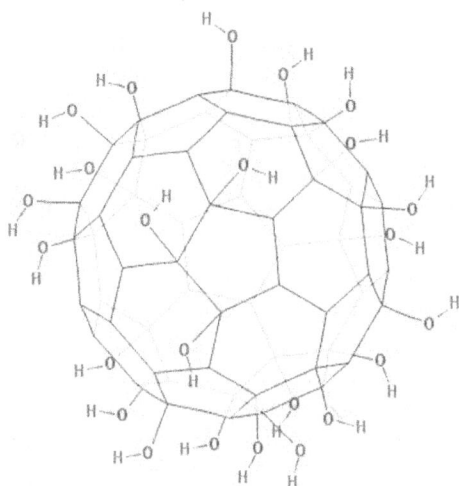

Figure 6.5. Example of a hydrophilic fullerene molecule [73]. This [Fullerenol A] image has been obtained by the author(s) from the Wikimedia website where it was made available by [Megotrend] under a CC BY-SA 4.0 licence. It is included within this chapter on that basis. It is attributed to [Megotrend].

these materials' toxicity is their high surface-to-volume ratio, ranking fullerene as the third most toxic material in their research; the first was graphene, the second was CNTs, and the last was carbon nanowires. Another important aspect highlighted by the authors was the oxidative stress induced by most of these nanomaterials. The ROS they can induce can damage proteins, membrane lipids, and DNA, potentially leading to apoptosis or cancer development. Furthermore, Madannejad *et al* mentioned that fullerenes might have hepatotoxicity but also documented their cardioprotective effects (Madannejad *et al* 2019).

Tao Jiang and coworkers conducted a study on the toxicity of carbon nano-materials, concluding that larger fullerenes (C_{70}, C_{84}) were less toxic than C_{60}, which is probably related to their smaller surface-to-volume ratios and their ability to penetrate the cell's interior. It was also observed that C_{60} interfered with the cell's pathway related to membrane integrity, while C_{70} and C_{84} did not lead to signs of cell wall damage (Jiang *et al* 2021). These observations were reinforced by the results from Nisoh and coworkers, who used visualizations from two different molecular dynamics methodologies to show that fullerenes tend to aggregate on the membrane at high concentrations, which might damage the cell wall (Nisoh *et al* 2022). Tao Jiang's research also concluded that DNA damage is a major mechanism of fullerene toxicity, as fullerenes can bind to DNA, and their small particles can reach the cell's nucleus. After exposure, an increased level of biomarkers related to DNA repair and antioxidant defenses can be found (Jiang *et al* 2021).

The hydrophobic surface of fullerenes, their optimal size, and the high versatility of their structure make them interesting candidates as nanodelivery tools (Youn *et al* 2017, Mohajeri *et al* 2019). In this regard, Minami and coworkers developed a fullerene that was complexed with siRNA (a type of RNA molecule that inhibits RNA translation) and plasma proteins to form particles in the range of micrometers. These particles were able to release siRNA into lung cells and suppress the expression of TLR4 gene resulting in reduced neutrophil accumulation after a challenge with intratracheal lipopolysaccharide (Minami *et al* 2014). Wang and coworkers created a C_{60} fullerene complexed with NH_2 and dextran, which presented an amphiphilic structure that formed aggregates similar to micelles; these were capable of protecting siRNA from being attacked by ROS and releasing the siRNA when exposed to visible light. As a consequence of this release, gene suppression rates of 53% and 69% were achieved for MDA-MB-231-EGFP cells and 4T1-GFP-Luc2 tumor-bearing mice, respectively (Wang *et al* 2017). Fullerenes are also vehicles capable of delivering DNA for transfection purposes. Nakamura and coworkers pioneered this field by functionalizing fullerenes with four positive charges that held DNA plasmid strands capable of transportation (Nakamura *et al* 2000, Kazemzadeh and Mozafari 2019).

Fullerenes have also been successfully applied to perform drug delivery. Paclitaxel, for example, is considered one of the most promising anticancer drugs; it can be successfully attached to fullerenes while still retaining its pharmacological activities (Gautam and Koshkina 2003, Zakharian *et al* 2005). Nanomaterials could also be used to attenuate the side effects of chemotherapy. For example, the collateral effects of doxorubicin were avoided when delivered through fullerene conjugation in a protocol to fight breast cancer. Normally, this drug would cause cardiomyopathy-related side effects, but these were decreased by using the nano-material, which offered 100% drug release at pH 5.25. Though the hydrophobicity of fullerene particles was a challenge, it was solved by using ethylene glycol spacers to improve water solubility (Kepinska *et al* 2018). This drug was also complexed with C82 fullerenes and cyclic arginyl-glycyl-aspartic acid peptide for delivery in lung cancer treatment. The structure presented high stability in phosphate solution, was well absorbed by non-small cell lung cancer cells, and was cytotoxic to these cells even at low concentrations (Zhao *et al* 2017, Mohajeri *et al* 2019). C_{60} nanoparticles could also be used for dermal and transdermal delivery. Inui and coworkers developed fullerenes complexed with alcohol groups capable of treating acne vulgaris. They applied the formulation to the sebaceous parts of mice and noted its ability to treat acne and its potential as a skincare material (Inui *et al* 2012, Garg and Jain 2022).

Fullerenes can also be used in bioimaging. A fluorescent C_{60} fullerene chemically modified to contain COOH groups on its surface was coated with mesoporous silica nanoparticles. The resulting material, a biocompatible water-soluble delivery vehicle that released doxorubicin hydrochloride in a pH-dependent manner, also allowed for cell imaging (Tan *et al* 2016). Another water-soluble fluorescent fullerene (C_{70}, which presents red fluorescence) was produced by Dreszer and coworkers and exhibited better bioavailability in *Drosophila melanogaster* when compared to nanoparticles containing the nanomaterial and pristine fullerene. The nanostructures proved to be biocompatible (Dreszer *et al* 2023). C_{20} fullerenes doped with the alkaline earth metal Ca were used to detect ethyl butyrate (a COVID-19 biomarker); a recovery time of 6.8 s was exhibited, making this system worthy of further attention (Ejaz *et al* 2024). DFT simulations showed that the fullerene C_{24} is capable of detecting the cancerous substance acrylamide, formed in starchy foods; this capability could be used to guarantee food safety. Spontaneous and reversible adsorption was predicted, making this material an interesting option for use in electrochemical sensors to detect the analyte (Tayebi-Moghaddam *et al* 2024). This type of simulation revealed that C_{24} fullerenes would exhibit better performance than Al12N12 and Al12P12 fullerenes when applied as part of a glucose sensor. The other two nanomaterials would be options to isolate the sugar from the mixture or serve as carriers (Tukadiya *et al* 2023). There are also sensors that use fullerenes to detect environmental contaminants; these nanomaterials can assist in the detection of H_2S and NO_2, with maximum sensitivities of 75.89% and 72.86%, respectively, at 200 °C (Naief *et al* 2023).

In the field of tissue engineering, fullerenes are also a promising nanomaterial. They can interact with cells, favoring proliferation and/or differentiation, for example. Hydrophilic self-assembled fullerene nanotubes and nanowhiskers inter-acted with vascular endothelial cells, offering, after seven days in 2D culture at 50 $\mu g\,ml^{-1}$, a proliferation of at least 400%. In a 3D scaffold of chitosan-based self-healing hydrogel, an increase in proliferation was also noted, advancing vascularization for tissue engineering applications (T Chen *et al* 2024). C_{60} fullerene impregnated with vancomycin was added to a nanocomposite of poly

(3-hydroxybutyrate-co-3-hydroxyvalerate) with nanohydroxyapatite to offer a biomaterial that could be applied in bone filling while also providing antibiotic therapy. The nanomaterial displayed flexibility, and the whole structure was biocompatible and capable of promoting macrophage cell adherence (Ribeiro *et al* 2023). Fullerenes may also support cutaneous healing. In one study, a biocompatible hydrogel containing fullerene C_{60} acted as ROS scavenger, showed antibacterial activity *in vitro*, and, in mice, accelerated wound closure. After seven days, treated wounds demonstrated 42.9% re-epithelialization, along with increased angiogenesis and collagen deposition (Chen *et al* 2023).

6.5 Biomedical applications of other important carbon nanomaterials

Besides graphene and its derivatives, CNTs and fullerenes, other nanomaterials based on carbon atoms also deserve attention due to their applications in the biomedical field. Carbon onions and CDs, among other carbon-based nanomaterials, for example, have already proven useful in biomedicine.

Concentric shells of carbon with a spherical shape form so-called carbon onions or carbon nano-onions: a zero-dimensional nanomaterial. These structures can be produced through methods that use nanodiamonds or graphite electrodes, for example (Cebik *et al* 2013, McDonough and Gogotsi 2013, Malode *et al* 2024): arc discharge, chemical vapor deposition, ion implantation, pyrolysis, and thermal annealing (Cabioc'h *et al* 2000, Chen *et al* 2001, Cabioc'h *et al* 2002, Palkar *et al* 2007, Echegoyen *et al* 2010, Garcia-Martin *et al* 2013). Carbon nano-onions are zero-dimensional and have sizes commonly ranging between 3 and 100 nm (Ahlawat *et al* 2021). They have already proven useful, for example, in drug delivery (where they last for a long time in the circulation, being biocompatible and presenting a large surface area to serve as a carrier) (Mamidi *et al* 2020); tissue engineering (carbon nano-onions associated with chitosan and polyvinyl alcohol formed films that were successfully subdermally implanted in Wistar rats, offering more stability and a good nano-composite reabsorption rate) (Grande Tovar *et al* 2020); bioimaging (the nano-material, emitting white light, can be modified with manganese oxide nanosheets to work as a biocompatible and water-soluble nanoprobe responsive to glutathione *in vivo* and *in vitro*) (Revuri *et al* 2018); sensing (a screen-printed electrode containing a chitosan matrix immobilizing carbon nano-onions associated with tyrosinase detected the herbicide glyphosate; the nanomaterial offered enhanced sensitivity) (Sok and Fragoso 2019); and as targeted structures (biocompatible nano-onions modified with hyaluronic acid and a phospholipid targeted cancer cells overexpressing CD44, being internalized by them) (D'Amora *et al* 2020).

CDs are particles formed from carbon atoms. Their size is less than 10 nm, and they have a quasi-spherical shape. They are soluble in water, biocompatible, and present photoluminescence that does not depend upon metal atoms. The sp^3 carbons on their surface allow functionalization to optimize the nanomaterial to perform specific tasks of interest, such as bioimaging and therapy. These nanostructures can be synthesized by applying pemetrexed carbon precursors that allow targeted association with folate receptors in breast cancer cells. Such nanostructures negatively impacted those cells and allowed observation through fluorescence emission (Noorkhajavi *et al* 2024). CDs have also successfully been used together with iron ions to produce $Fe_5H_9O_{15}/Fe_2O_3@sp^2$-carbon colloidal nanoparticles to build a low-cost voltammetric dopamine monitor that offered selective detection of this substance in urine (Guye *et al* 2024). Therefore, carbon-based nanomaterials have a large potential still to be explored in terms of different kinds of applications in the biomedical field.

6.6 Green methods for producing eco-friendly carbon nanomaterials

This chapter has demonstrated that carbon nanomaterials present a vast array of important uses in the biomedical field, offering efficiency. However, especially for applications in biological systems, efficiency is not the only factor that needs to receive attention. Today, there is a necessity to offer sustainability and biocompatibility through eco-friendly protocols and tools. As previously mentioned, environmental pollution is a problem worldwide; anthropogenic actions have caused damage to ecosystems, representing a threat to their balance and to the survival of living beings. Consequently, it is desirable that the carbon nanomaterials applied in biomedicine are produced through green protocols, avoiding damage to the environment and undesirable toxicity when applied in clinical use. Eco-friendly protocols commonly produce less cytotoxic nanomaterials, which are of special interest in the biomedical field.

The conventional methods used to produce carbon nanomaterials, in contrast to green methods, normally require large amounts of energy and raw materials and/or the use of toxic chemicals, resulting in the generation of dangerous pollutants. Eco-friendly methods are also efficient, but they are sustainable and commonly present lower costs. The raw materials used come from biological sources such as plants, bacteria, fungi, and algae, or products generated by these organisms. Honey, for example, can already be used to produce luminescent carbon nanoparticles in an eco-friendly manner. In one study, luminescent carbon nanoparticles with rapid clearance properties were produced. These structures, approximately 7 nm in size, were used in the immediate imaging of sentinel lymph nodes. A rapid signal was produced, optimizing the efficiency of real-time high-resolution intraoperative photoacoustic imaging in conjunction with a near-infrared probe (Wu *et al* 2013).

Biomass from the agroindustry is also an interesting raw material for the synthesis of carbon-based nanomaterials, such as nanoparticles to be applied in the biomedical field (Qasim *et al* 2023). Different types of protocols can be applied to process the biomass, and biological protocols, using living systems, are among them (McKendry 2002). A large array of existing cost-effective and sustainable protocols to produce carbon-based nanomaterials offers various possibilities for final products, such as CNTs (Ghosh *et al* 2007), carbon quantum dots (Lu *et al* 2012), and graphene and its derivatives (Kim *et al* 2011, Ruan *et al* 2011, Bose *et al* 2012). Nanomaterials can be produced in the form of optimized tools that perform a desirable task and/or allow specific conditions during synthesis (such as temperature). Bacteria and other microorganisms can be used in the synthesis of these nanomaterials (Moradi *et al* 2021). For example, *Staphylococcus aureus* was used to produce blue–green luminescent carbon dots rapidly at low temperatures. The nanoscale material produced offered efficient imaging and a strong antibacterial effect (Zhao *et al* 2021). So, is it possible to adopt a green protocol when the aim is to produce carbon-based nanomaterials, dedicating attention to sustainability principles?

6.7 Conclusions and future perspectives

Carbon-based nanomaterials have already been shown to have the potential to be applied in different areas of biomedicine, such as drug delivery, therapy, sensors, bioimaging, and tissue engineering. However, some important aspects, such as biocompatibility, still need additional attention. For example, although graphene can be part of an efficient drug delivery system, some reports suggest this material can cause oxidative stress and clustering in tissues (Malode *et al* 2024). Therefore, it is expected that attention will continue to be dedicated to developing cost-effective, efficient, biocompatible, and sustainable carbon-based nanomaterials to precisely perform roles in biomedicine. New and innovative protocols still need to be proposed, and the nanomaterials produced need to continue to be assayed not only *in vitro* but also *in vivo* to guarantee safety in future clinical translation. Different nanomaterials present different general characteristics and behaviors. For example, nanodiamonds are commonly more biocompatible than fullerenes and CNTs (Zhang *et al* 2011); however, functionalization can mitigate undesirable effects and optimize desirable characteristics to improve efficiency. Nanotechnology, in particular carbon-based nanomaterials, promises to offer strategic tools to address difficulties faced in the biomedical field.

References and further reading

Acharya R, Patil T V, Dutta S D, Lee J, Ganguly K, Kim H, Randhawa A and Lim K T 2024 Single-walled carbon nanotube-based optical nano/biosensors for biomedical applications: role in bioimaging, disease diagnosis, and biomarkers detection *Adv. Mater. Technol. (online first)* 2400279

Ahlawat J, Masoudi Asil S, Guillama Barroso G, Nurunnabi M and Narayan M 2021 Application of carbon nano onions in the biomedical field: recent advances and challenges *Biomater. Sci.* **9** 626–44

Ajeel F N, Bardan K H, Kareem S H and Khudhair A M 2023 Pd doped carbon nanotubes as a drug carrier for Gemcitabine anticancer drug: DFT studies *Chem. Phys. Impact* **7** 100298

Bacon M, Bradley S J and Nann T 2014 Graphene quantum dots *Part. Part. Syst. Charact.* **31** 415–28

Badea N, Craciun M M, Dragomir A S, Balas M, Dinischiotu A, Nistor C, Gavan C and Ionita D 2020 Systems based on carbon nanotubes with potential in cancer therapy *Mater. Chem. Phys.* **241** 122435

Barhoum A, Shalan A E, El-Hout S I, Ali G A M, Abdelbasir S M, Serea E S A, Ibrahim A H and Pal K 2019 A broad family of carbon nanomaterials: classification, properties, synthesis, and emerging applications ed A Barhoum, M Bechelany and A Makhlouf *Handbook of Nanofibers* (Cham: Springer)

Biru E I, Necolau M I, Zainea A and Iovu H 2022 Graphene oxide–protein-based scaffolds for tissue engineering: recent advances and applications *Polymers* **14** 1032

Boddu A, Obireddy S R, Zhang D, Rao K K and Lai W F 2022 ROS-generating, pH-responsive and highly tunable reduced graphene oxide-embedded microbeads showing intrinsic anti-cancer properties and multi-drug co-delivery capacity for combination cancer therapy *Drug Deliv.* **29** 2481–90

Bogdanović G and Djordjević A 2016 Carbon nanomaterials: biologically active fullerene derivatives *Srp Arh Celok Lek* **144** 222–31

Bose S, Kuila T, Mishra A K, Kimd N H and Lee J H 2012 Dual role of glycine as a chemical functionalizer and a reducing agent in the preparation of graphene: an environmentally friendly method *J. Mater. Chem.* **22** 9696

Bosi S, Ballerini L and Prato M 2014 Carbon nanotubes in tissue engineering *Top. Curr. Chem.* **348** 181–204

Cao X T, Patil M P, Phan Q T, Le C M, Ahn B H, Kim G D and Lim K T 2020 Green and direct functionalization of poly (ethylene glycol) grafted polymers onto single-walled carbon nanotubes: effective nanocarrier for doxorubicin delivery *J. Ind. Eng. Chem.* **83** 173–80

Cabioc'h T, Jaouen M, Thune E, Guerin P, Fayoux C and Denanot M 2000 Carbon onions formation by high-dose carbon ion implantation into copper and silver *Surf. Coat. Technol.* **128** 43–50

Cabioc'h T, Thune E, Riviere J, Camelio S, Girard J, Guerin P, Jaouen M, Henrard L and Lambin P 2002 Structure and properties of carbon onion layers deposited onto various substrates *J. Appl. Phys.* **91** 1560–7

Cebik J, McDonough J K, Peerally F, Medrano R, Neitzel I, Gogotsi Y and Osswald S 2013 Raman spectroscopy study of the nanodiamond-to-carbon onion transformation *Nanotechnology* **24** 205703

Chen T Y, Cheng K C, Yang P S, Shrestha L K, Ariga K and Hsu S 2024 Interaction of vascular endothelial cells with hydrophilic fullerene nanoarchitectured structures in 2D and 3D environments *Sci. Technol. Adv. Mater.* **25** 2315014

Chen X, Deng F, Wang J, Yang H, Wu G, Zhang X, Peng J and Li W 2001 New method of carbon onion growth by radio-frequency plasma-enhanced chemical vapor deposition *Chem. Phys. Lett.* **336** 201–4

Chen X, Zhang Y, Yu W, Zhang W, Tang H and Yuan W E 2023 In situ forming ROS-scavenging hybrid hydrogel loaded with polydopamine-modified fullerene nanocomposites for promoting skin wound healing *J. Nanobiotechnol.* **21** 129

Chen Y L, Kurniawan D, Tsai M D, Chang J W, Chang Y N, Yang S C, Chiang W H and Kung C W 2024 Two-dimensional metal–organic framework for post-synthetic immobilization of graphene quantum dots for photoluminescent sensing *Commun. Chem.* **7** 108

D'Amora M, Camisasca A, Boarino A, Arpicco S and Giordani S 2020 Supramolecular functionalization of carbon nano-onions with hyaluronic acid-phospholipid conjugates for selective targeting of cancer cells *Colloids Surf. B* **188** 110779

Delogu L G *et al* 2012 Functionalized multiwalled carbon nanotubes as ultrasound contrast agents *Proc. Natl. Acad. Sci. USA* **109** 16612–7

Dreszer D, Szewczyk G, Szubka M, Maroń A M, Urbisz A Z, Małota K, Sznajder J, Rost-Roszkowska M, Musioł R and Serda M 2023 Uncovering nanotoxicity of a water-soluble and red-fluorescent fullerene nanomaterial *Sci. Total Environ.* **879** 163052

Ebbesen T W 1994 Carbon nanotubes *Annu. Rev. Mater. Sci.* **24** 235–64

Echegoyen L, Ortiz A, Chaur M N and Palkar A J 2010 Carbon Nano Onions *Chemistry of Nanocarbons* ed T Akasaka, F Wudl and S Nagase (New York: Wiley) 463–83

Ejaz M, AlMohamadi H, Khan A L, Yasin M, Mahmood T, Ayub K, Tabassum S and Gilani M A 2024 A rational design of metal-doped C20 fullerene-based sensor for the selective detection of ethyl butyrate as COVID-19 biomarker *Surf. Interfaces* **52** 104869

Esmaeili Y, Bidram E, Zarrabi A, Amini A and Cheng C 2020 Graphene oxide and its derivatives as promising *in vitro* bio-imaging platforms *Sci. Rep.* **10** 18052

Felegari Z and Hamedani S 2022 Adsorption properties and quantum molecular descriptors of the anticancer drug cytophosphane on the armchair single-walled carbon nanotubes: a DFT study *Lett. Org. Chem.* **19** 1034–41

Ferreira G M D, Ferreira G M D, Franca J R and Soares J R 2022 Carbonaceous materials for nanoremediation of polluted and nutrient-depleted soils *Nanotechnology for Environmental Pollution Decontamination: Tools, Methods, and Approaches for Detection and Remediation* ed F M P Tonelli, R A Bhat and G H Dar (New York: CRC Press) 64

Friedman S H, DeCamp D L, Sijbesma R P, Srdanov G, Wudl F and Kenyon G L 1993 Inhibition of the HIV-1 protease by fullerene derivatives: model building studies and experimental verification *J. Am. Chem. Soc.* **115** 6506–9

Garcia-Martin T, Rincon-Arevalo P and Campos-Martin G 2013 Method to obtain carbon nano-onions by pyrolisys of propane *Cent. Eur. J. Phys.* **11** 1548–58

Garg U and Jain K 2022 Dermal and transdermal drug delivery through vesicles and particles: preparation and applications *Adv. Pharm. Bull.* **12** 45

Gaur M, Misra C, Yadav A B, Swaroop S, Maolmhuaidh F Ó, Bechelany M and Barhoum A 2021 Biomedical applications of carbon nanomaterials: fullerenes, quantum dots, nanotubes, nanofibers, and graphene *Materials* **14** 5978

Gautam A and Koshkina N 2003 Paclitaxel (taxol) and taxoid derivatives for lung cancer treatment: potential for aerosol delivery *Curr. Cancer Drug Targets* **3** 287–96

Ghosh P, Afre R A, Soga T and Jimbo T 2007 A simple method of producing single-walled carbon nanotubes from a natural precursor: eucalyptus oil *Mater. Lett.* **61** 3768–70

Grande Tovar C D, Castro J I, Valencia C H, Navia Porras D P, Herminsul Mina Hernandez J, Valencia Zapata M E and Chaur M N 2020 Nanocomposite films of chitosan-grafted carbon nano-onions for biomedical applications *Molecules* **25** 1203

Gulati P, Mishra P and Islam S S 2020 Sensitive biosensor for chronic myeloid leukemia detection using multi-wall carbon nanotubes *AIP Conf. Proc.* (New York, NY: AIP Publishing) 2276

Guye M E, Appiah-Ntiamoah R, Dabaro M D and Kim H 2024 Engineering FeOOH/Fe2O3@carbon interfaces with biomass-derived carbon nanodot/iron colloids for efficient redox-modulated dopamine voltammetric detection *Chemistry* **19** e202400435

Han S, Liu W, Zheng M and Wang R 2020 Label-free and ultrasensitive electrochemical DNA biosensor based on urchinlike carbon nanotube-gold nanoparticle nanoclusters *Anal. Chem.* **92** 4780–7

Heidarian M, Khazaei A and Saien J 2021 Grafting drugs to functionalized single-wall carbon nanotubes as a potential method for drug delivery *Phys. Chem. Res.* **9** 57–68

Henna T and Pramod K 2020 Graphene quantum dots redefine nanobiomedicine *Mater. Sci. Eng.* C **110** 110651

Hu J, Liu J, Gan L and Long M 2019 Surface-modified graphene oxide-based cotton fabric by ion implantation for enhancing antibacterial activity *ACS Sust. Chem Eng* **7** 7686–92

Hu Y and Niemeyer C M 2020 Designer DNA–silica/carbon nanotube nanocomposites for traceable and targeted drug delivery *J. Mater. Chem.* B **8** 2250–5

Huang K, Liu X, Lv Z, Zhang D, Zhou Y, Lin Z and Guo J 2023 MMP9-responsive graphene oxide quantum dot-based nano-in-micro drug delivery system for combinatorial therapy of choroidal neovascularization *Small* **19** e2207335

Huang X, Zhang J, Zhang L, Su H, Liu X and Liu J 2020 A sensitive H_2O_2 biosensor based on carbon nanotubes/tetrathiafulvalene and its application in detecting NADH *Anal. Biochem.* **589** 113493

Huston M, DiBella M and Gupta A 2021 Green synthesis of nanomaterials *Nanomaterials* **11** 2130

Inui S, Aoshima H, Ito M, Kobuko K and Itami S 2012 Inhibition of sebum production and Propionibacterium acnes lipase activity by fullerenol, a novel polyhydroxylated fullerene: potential as a therapeutic reagent for acne *J. Cosmet. Sci.* **63** 259–65

Jain R, Dhanjai and Sinha A 2016 Graphene-zinc oxide nanorods nanocomposite based sensor for voltammetric quantification of tizanidine in solubilized system *Appl. Surf. Sci.* **369** 151–8

Jiang T, Lin Y, Amadei C A, Gou N, Rahman S M, Lan J, Vecitis C D and Gu A Z 2021 Comparative and mechanistic toxicity assessment of structure-dependent toxicity of carbon-based nanomaterials *J. Hazard. Mater.* **418** 126282

Jiřičková A, Jankovský O, Sofer Z and Sedmidubský D 2022 Synthesis and applications of graphene oxide *Materials* **15** 920

Kang S, Kim K M, Jung K, Son Y, Mhin S, Ryu J H, Shim K B, Lee B, Han H and Song T 2019 Graphene oxide quantum dots derived from coal for bioimaging: facile and green approach *Sci. Rep.* **9** 4101

Karachi N 2023 Computational investigation on anti-cancer drugs/carbon nanotube as a drug delivery system *J. Optoelectron. Nanostruct.* **8** 1–17 https://www.magiran.com/p2675917

Karimi M *et al* 2015 Carbon nanotubes part I: preparation of a novel and versatile drug-delivery vehicle *Exp. Opin. Drug Deliv.* **12** 1071–87

Kazemzadeh H and Mozafari M 2019 Fullerene-based delivery systems *Drug Discov. Today* **24** 898–905

Kepinska M, Kizek R and Milnerowicz H 2018 Fullerene as a doxorubicin nanotransporter for targeted breast cancer therapy: capillary electrophoresis analysis *Electrophoresis* **39** 2370–9

Khakpour E, Salehi S, Naghib S M, Ghorbanzadeh S and Zhang W 2023 Graphene-based nanomaterials for stimuli-sensitive controlled delivery of therapeutic molecules *Front. Bioeng. Biotechnol.* **11** 1129768

Kim Y K, Kim M H and Min D H 2011 Biocompatible reduced graphene oxide prepared by using dextran as a multifunctional reducing agent *Chem. Commun.* **47** 3195–7

Kumar A, Sharma K and Dixit A R 2020 Role of graphene in biosensor and protective textile against viruses *Med. Hypotheses* **144** 110253

Biddulph M 2007 Buckminster Fuller biosphere *[Photograph]* (Flickr) https://www.flickr.com/photos/mbiddulph/558711477

Li W, Gao Y, Zhang J, Wang X, Yin F, Li Z and Zhang M 2020 Universal DNA detection realized by peptide based carbon nanotube biosensors *Nanoscale Adv.* **2** 717–23

Lin J, Lin J H, Yeh T Y, Zheng J H, Cho E C and Lee K C 2024 Fabrication of hyaluronic acid with graphene quantum dot as a dual drug delivery system for cancer therapy *FlatChem.* **44** 100607

Lin Z, Wu G, Zhao L and Lai K W C 2019 Carbon nanomaterial-based biosensors: a review of design and applications *IEEE Nanatechnol. Mag.* **13** 4–14

Liu Z, Li X, Tabakman S M, Jiang K, Fan S and Dai H 2008 Multiplexed multicolor Raman imaging of live cells with isotopically modified single-walled carbon nanotubes *J. Am. Chem. Soc.* **130** 13540–1

Lourenço V O, Silva C S, Tonelli F M P, Santinelli B, Pinto G C, Paixão B, Prote L C S, Delgado V M S and Tonelli F C P 2024 Main biological contaminants endangering humans' health *Nanotechnology-Based Sensors for Detection of Environmental Pollution* 1st edn ed F M P Tonelli, A Roy, M Ozturk and H C A Murthy (New York: Elsevier) pp 1–26

Lu W, Qin X, Liu S, Chang G, Zhang Y, Luo Y, Asiri A M, Al-Youbi A O and Sun X 2012 Economical, green synthesis of fluorescent carbon nanoparticles and their use as probes for sensitive and selective detection of mercury(II) ions *Anal. Chem.* **84** 5351–7

Madannejad R, Shoaie N, Jahanpeyma F, Darvishi M H, Azimzadeh M and Javadi H 2019 Toxicity of carbon-based nanomaterials: reviewing recent reports in medical and biological systems *Chem. Biol. Interact.* **307** 206–22

Maiti D, Tong X, Mou X and Yang K 2019 Carbon-based nanomaterials for biomedical applications: a recent study *Front. Pharmacol.* **9** 1401

Malode S J, Pandiaraj S, Alodhayb A and Shetti N P 2024 Carbon nanomaterials for biomedical applications: progress and outlook publication *ACS Appl. Bio Mater.* **7** 752–77

Mamidi N, Zuníga A E and Villela-Castrejón J 2020 Engineering and evaluation of forcespun functionalized carbon nano-onions reinforced poly (ε-caprolactone) composite nanofibers for pH-responsive drug release *Mater. Sci. Eng.* C **112** 110928

Mansha S, Sajjad A, Zarbab A, Afzal T, Kanwal Z, Iqbal M J, Raza M A and Ali S 2024 Development of pH-responsive, thermosensitive, antibacterial, and anticancer CS/PVA/ graphene blended hydrogels for controlled drug delivery *Gels* **10** 205

Mauri E, Salvati A, Cataldo A, Mozetic P, Basoli F, Abbruzzese F, Trombetta M, Bellucci S and Rainer A 2021 Graphene-laden hydrogels: a strategy for thermally triggered drug delivery *Mater. Sci. Eng.* C **118** 111353

McDonough J K and Gogotsi Y 2013 Carbon onions: synthesis and electrochemical applications *Electrochem. Soc. Interface* **22** 61–6

McKendry P 2002 Energy production from biomass (Part 2): conversion technologies *Bioresour. Technol.* **83** 47–54

Megotrend 2014 File:Fullerenol A.png https://commons.wikimedia.org/wiki/File:Fullerenol_% D0%90.png via Wikimedia Commons

Minami K, Okamoto K, Doi K, Harano K, Noiri E and Nakamura E 2014 siRNA delivery targeting to the lung via agglutination induced accumulation and clearance of cationic tetraamino fullerene *Sci. Rep* **4** 4916

Minsu P, Hyewon Y and Seokwoo J 2020 Graphene-based quantum dot emitters for light-emitting diodes *Graphene for Flexible Lighting and Displays* ed T W Lee (New York: Elsevier) 117–50

Mohajeri M, Behnam B and Sahebkar A 2019 Biomedical applications of carbon nanomaterials: drug and gene delivery potentials *J. Cell. Physiol.* **234** 298–319

Mohammadpour S, Mokhtarzade A, Jafari-Ramiani A and Solati-Hashjin M 2024 Tailoring surface properties of poly(caprolactone)/hydroxyapatite scaffolds through aminolysis and multi-walled carbon nanotube coating for bone tissue engineering *Mater. Today Commun.* **39** 109056

Moradi M, Molaei R, Kousheh S A, Guimarães J T and McClements D J 2021 Carbon dots synthesized from microorganisms and food by-products: active and smart food packaging applications *Crit. Rev. Food Sci. Nutr.* **2021** 1–17

Moura P C, Pivetta T P, Vassilenko V, Ribeiro P A and Raposo M 2023a Graphene oxide thin films for detection and quantification of industrially relevant alcohols and acetic acid *Sensors* **23** 462

Moura P C, Ribeiro P A, Raposo M and Vassilenko V 2023b The state of the art on graphene-based sensors for human health monitoring through breath biomarkers *Sensors* **23** 9271

Murjani B O, Kadu P S, Bansod M, Vaidya S S and Yadav M D 2022 Carbon nanotubes in biomedical applications: current status, promises, and challenges *Carbon Lett.* **32** 1207–26

Nagai Y, Hamano R, Nakamura K, Widjaja I A, Tanaka N, Zhang M, Tanaka T, Kataura H, Yudasaka M and Fujigaya T 2024 Bright NIR-II fluorescence from biocompatible gel-coated carbon nanotubes for *in vivo* imaging *Carbon* **218** 118728

Naief M F, Mohammed S N, Ahmed Y N, Mohammed A M, Mohammed S N and Mohammed S N 2023 Novel preparation method of fullerene and its ability to detect H_2S and NO_2 gases *Res. Chem.* **5** 100924

Nakamura E, Tokuyama H, Yamago S, Shiraki T and Sugiura Y 1996 Biological activity of water-soluble fullerenes. Structural dependence of dna cleavage, cytotoxicity, and enzyme inhibitory activities including HIV-protease inhibition *Bull. Chem. Soc. Jpn.* **69** 2143–51

Nakamura E, Isobe H, Tomita N, Sawamura M, Jinno S and Okayama H 2000 Functionalized fullerene as an artificial vector for transfection *Angew. Chem.* **112** 4424–7

Neri G, Leonardi S G, Latino M, Donato N, Baek S, Conte D E, Russo P A and Pinna N 2013 Sensing behavior of SnO_2/reduced graphene oxide nanocomposites toward NO_2 *Sensors Actuators* B **179** 61–8

Nikeafshar S, Khazaei A and Tahvilian R 2022 Inhibition of methamphetamine-induced cytotoxicity in the U87-cell line by atorvastatin-conjugated carbon nanotubes *Appl. Biochem. Biotechnol.* **2022** 1–25

Nisoh N, Jarerattanachat V, Karttunen M and Wong-Ekkabut J 2022 Fullerenes' interactions with plasma membranes: insight from the MD simulations *Biomolecules* **12** 639

Noor N, Mutalik S, Younas M W, Chan C Y, Thakur S, Wang F *et al* 2019 Durable antimicrobial behaviour from silver-graphene coated medical textile composites *Polymers* **11** 2000

Noorkhajavi G, Abdian N, Najaflou M, Hefferon K, Yari-Khosroushahi A and Shahgolzari M 2024 Synthesis of self-targeted carbon nanodots for efficient cancer cell imaging and therapy *Inorg. Chem. Commun.* **161** 112027

Norizan M N, Moklis M H, Demon S Z N, Halim N A, Samsuri A, Mohamad I S, Knight V F and Abdullah N 2020 Carbon nanotubes: functionalisation and their application in chemical sensors *RSC Adv.* **10** 43704–32

Notarianni M, Liu J, Vernon K and Motta N 2016 Synthesis and applications of carbon nanomaterials for energy generation and storage *Beilstein J. Nanotechnol.* **7** 149–96

Nováček M, Jankovský O, Luxa J, Sedmidubský D, Pumera M, Fila V, Lhotka M, Klímová K, Matějková S and Sofer Z 2017 Tuning of graphene oxide composition by multiple oxidations for carbon dioxide storage and capture of toxic metals *J. Mater. Chem.* A **5** 2739–48

Novir S B and Aram M R 2021 Quantum mechanical studies of the adsorption of Remdesivir, as an effective drug for treatment of COVID-19, on the surface of pristine, COOH-functionalized and S-, Si-, and Al-doped carbon nanotubes *Phys.* E **129** 114668

Novoselov K S, Geim A K, Morozov S, Dubonos S, Zhang Y and Jiang D 2004 Room-temperature electric field effect and carrier-type inversion in graphene films arXiv Preprint https://arxiv.org/abs/cond-mat/0410631

Palkar A, Melin F, Cardona C M, Elliott B, Naskar A K, Edie D D, Kumbhar A and Echegoyen L 2007 Reactivity differences between carbon nano onions (CNOs) prepared by different methods *Chem.-Asian J.* **2** 625–33

Phelane L, Gouveia-Caridade C, Barsan M M, Baker P G, Brett C M and Iwuoha E I 2020 Electrochemical determination of tyrosine using a novel tyrosinase multi-walled carbon nanotube (MWCNT) polysulfone modified glassy carbon electrode (GCE) *Anal. Lett.* **53** 308–21

Qasim M, Clarkson A N and Hinkley S F R 2023 Green synthesis of carbon nanoparticles (CNPs) from biomass for biomedical applications *Int. J. Mol. Sci.* **24** 1023

Rahmani E, Pourmadadi M, Zandi N, Rahdar A and Baino F 2022 pH-responsive PVA-based nanofibers containing GO modified with Ag nanoparticles: physico-chemical characterization, wound dressing, and drug delivery *Micromachines* **13** 1847

Rajaei M, Rashedi H, Yazdian F, Navaei-Nigjeh M, Rahdar A and Díez-Pascual A M 2023 Chitosan/agarose/graphene oxide nanohydrogel as drug delivery system of 5-fluorouracil in breast cancer therapy *J. Drug Deliv. Sci. Technol.* **82** 104307

Resende R R, Fonseca E A, Tonelli F M P, Sousa B R, Santos A K, Gomes K N, Guatimosim S C, Kihara A H and Ladeira L O 2014 Scale/topography of substrates surface resembling extracellular matrix for tissue engineering *J. Biomed. Nanotechnol.* **10** 1157–93

Revuri V, Cherukula K, Nafiujjaman M, Jae K C, Park I K and Lee Y K 2018 White-light-emitting carbon nano-onions: a tunable multichannel fluorescent nanoprobe for glutathione-responsive bioimaging *ACS Appl. Nano Mater.* **1** 662–74

Ribeiro M E A, Huaman N R C, Folly M M, Gomez J G C and Rodríguez R J S 2023 A potential hybrid nanocomposite of poly(3-hydroxybutyrate-co-3-hydroxyvalerate) and fullerene for bone tissue regeneration and sustained drug release against bone infections *Int. J. Biol. Macromol.* **251** 126531

Riley P R and Narayan R J 2021 Recent advances in carbon nanomaterials for biomedical applications: a review *Curr. Opin. Biomed. Eng.* **17** 100262

Ruan G, Sun Z, Peng Z and Tour J M 2011 Growth of graphene from food, insects, and waste *ACS Nano* **5** 7601–7

Sahoo D, Mitra T, Chakraborty K and Sarkar P 2022 Remotely controlled electro-responsive on-demand nanotherapy based on amine-modified graphene oxide for synergistic dual drug delivery *Mater. Today Chem* **25** 100987

Samadian H, Mohammad-Rezaei R, Jahanban-Esfahlan R, Massoumi B, Abbasian M, Jafarizad A and Jaymand M 2020 A de novo theranostic nanomedicine composed of PEGylated graphene oxide and gold nanoparticles for cancer therapy *J. Mater. Res.* **35** 430–41

Seo G, Lee G, Kim M J, Baek S H, Choi M and Ku K B 2020 Rapid detection of COVID-19 causative virus (SARS-CoV-2) in human nasopharyngeal swab specimens using field-effect transistor-based biosensor *ACS Nano* **14** 5135–42

Shi J, Votruba A R, Farokhzad O C and Langer R 2010 Nanotechnology in drug delivery and tissue engineering: from discovery to applications *Nano Lett.* **10** 3223–30

Shu Y, Lu Q, Yuan F, Tao Q, Jin D, Yao H, Xu Q and Hu X 2020 Stretchable electrochemical biosensing platform based on Ni-MOF composite/Au nanoparticle-coated carbon nanotubes for real-time monitoring of dopamine released from living cells *ACS Appl. Mater. Interfaces* **12** 49480–8

Silva C S *et al* 2024 Nanoremediation and antioxidant potential of biogenic silver nanoparticles synthesized using leucena's leaves, stem, and fruits *Int. J. Mol. Sci.* **25** 3993

Sima L E, Chiritoiu G, Negut I, Grumezescu V, Orobeti S, Munteanu C V A, Sima F and Axente E 2020 Functionalized graphene oxide thin films for anti-tumor drug delivery to melanoma cells *Front Chem.* **8** 184

Sinha A, Dhanjai and Jain R 2015 Electrocatalytic determination of α 2-adrenergic agonist tizanidine at graphene–silicon dioxide nanocomposite sensor *Mater. Res. Bull.* **65** 307–14

Sok V and Fragoso A 2019 Amperometric biosensor for glyphosate based on the inhibition of tyrosinase conjugated to carbon nano-onions in a chitosan matrix on a screen-printed electrode *Microchim. Acta* **186** 569

Sontakke A D, Tiwari S, Gupta P, Banerjee S K and Purkait M K 2024 Room temperature synthesis of β-cyclodextrin functionalized graphene oxide decorated MIL-100 (Fe): a sustainable drug cargo for anticancer drug delivery *Mater. Today Commun.* **38** 108560

Sun C, Xie Y, Zhu H, Zheng X, Hou R, Shi Z, Li J and Yang Q 2023 Highly electroactive tissue engineering scaffolds based on nanocellulose/sulfonated carbon nanotube composite hydrogels for myocardial tissue repair *Biomacromolecules* **24** 5989–97

Tan L, Wu T, Tang Z W, Xiao J Y, Zhuo R X, Shi B and Liu C J 2016 Water-soluble photoluminescent fullerene capped mesoporous silica for pH-responsive drug delivery and bioimaging *Nanotechnology* **27** 315104

Tayebi-Moghaddam S, Aliakbari M and Tayeboun K 2024 Fullerene (C_{24}) as a potential sensor for the detection of acrylamide: a DFT study *Int. J. New Chem.* **11** 82–9

Thang P Q 2020 A green process of polyampholyte-grafted single-walled carbon nanotubes for enhanced anticancer drug delivery *(Doctoral Dissertation)* https://repository.pknu.ac.kr:8443/handle/2021.oak/23672

Tonelli F M P, Delgado V M S, Lourenço V O, Silva C S, Pinto G C, Santinelli B and Tonelli F C P 2024 Main organic pollutants and their risk to living beings *Nanotechnology-Based Sensors for Detection of Environmental Pollution* 1st ed ed F M P Tonelli, A Roy, M Ozturk and H C A Murthy (New York: Elsevier)

Tonelli F M P and Tonelli F C P 2020 Role of modern innovative techniques for assessing and monitoring heavy metal and pesticide pollution in different environments *Bioremediation and Biotechnology* **vol 2** 1st edn ed R A Bhat, K R Hakeem and M A Dervash (Berlin: Springer) 25–37

Tonelli F M P, Goulart V A, Gomes K N, Ladeira M S, Santos A K, Lorençon E, Ladeira L O and Resende R R 2015 Graphene-based nanomaterials: biological and medical applications and toxicity *Nanomedicine* **10** 2423–50

Tonelli F M P, Santos A K, Gomes K N, Lorençon E, Guatimosin S, Ladeira L O and Resende R R 2012 Carbon nanotube interaction with extracellular matrix proteins producing scaffolds for tissue engineering *Int. J. Nanomed.* **2012** 4511–29

Tukadiya N A, Jana S K, Chakraborty B and Jha P K 2023 C24 Fullerene and its derivatives as a viable glucose sensor: DFT and TD-DFT studies *Surf. Interfaces* **41** 103220

Tzirakis M D and Orfanopoulos M 2013 Radical reactions of fullerenes: from synthetic organic chemistry to materials science and biology *Chem. Rev.* **113** 5262–321

Wang J, Xie L, Wang T, Wu F, Meng J, Liu J and Xu H 2017 Visible light-switched cytosol release of siRNA by amphiphilic fullerene derivative to enhance RNAi efficacy *in vitro* and *in vivo Acta Biomater.* **59** 158–69

Wang X, Guo W, Li L, Yu F, Li J, Liu L, Fang B and Xia L 2021 Photothermally triggered biomimetic drug delivery of Teriparatide via reduced graphene oxide loaded chitosan hydrogel for osteoporotic bone regeneration *Chem. Eng. J.* **413** 127413

Wang X, Zhu Y, Chen M, Yan M, Zeng G and Huang D 2019 How do proteins 'response' to common carbon nanomaterials? *Adv. Colloid Interface Sci.* **270** 101–7

Wang Y, Li Z, Wang J, Li J and Lin Y 2011 Graphene and graphene oxide: biofunctionalization and applications in biotechnology *Trends Biotechnol.* **29** 205–12

Wu J, Qin Z, Jiang X, Fang D, Lu Z, Zheng L and Zhao J 2022 ROS-responsive PPGF nanofiber membrane as a drug delivery system for long-term drug release in attenuation of osteo-arthritis *NPJ Regen. Med.* **7** 66

Wu L *et al* 2013 A green synthesis of carbon nanoparticles from honey and their use in real-time photoacoustic imaging *NanoResearch* **6** 312–25

Xia Y, Li S, Nie C, Zhang J, Zhou S, Yang H, Li M, Li W, Cheng C and Haag R 2019 A multivalent polyanion-dispersed carbon nanotube toward highly bioactive nanostructured fibrous stem cell scaffolds *Appl. Mater. Today* **16** 518–28

Xu G, Wang J, Si G, Wang M, Cheng H and Chen B 2017 Preparation, photoluminescence properties and application for *in vivo* tumor imaging of curcumin derivative-functionalized graphene oxide composite *Dyes Pigments* **141** 470–8

Yang K, Feng L and Liu Z 2015 The advancing uses of nano-graphene in drug delivery *Expert Opin. Drug Deliv.* **12** 601–12

Yang M, Wang H, Liu P and Cheng J 2021 A 3D electrochemical biosensor based on super-aligned carbon nanotube array for point-of-care uric acid monitoring *Biosens. Bioelectron.* **179** 113082

Yang N, Chen X, Ren T, Zhang P and Yang D 2015 Carbon nanotube based biosensors *Sens. Actuators* B **207** 690–715

Yang X X, Li C M, Li Y F, Wang J and Huang C Z 2017 Synergistic antiviral effect of curcumin functionalized graphene oxide against respiratory syncytial virus infection *Nanoscale* **9** 16086–92

Youn Y S, Kwag D S and Lee E S 2017 Multifunctional nano-sized fullerenes for advanced tumor therapy *J. Pharm. Invest.* **47** 1–10

Yudasaka M *et al* 2017 Near-infrared photoluminescent carbon nanotubes for imaging of brown fat *Sci. Rep.* **7** 44770

Zainal-Abidin M H, Hayyan M, Ngoh G C, Wong W F and Looi C Y 2020 Potentiating the anti-cancer profile of tamoxifen-loaded graphene using deep eutectic solvents as functionalizing agents *Appl. Nanosci.* **10** 293–304

Zakharian T Y, Seryshev A, Sitharaman B, Gilbert B E, Knight V and Wilson L J 2005 A fullerene-paclitaxel chemotherapeutic: synthesis, characterization, and study of biological activity in tissue culture *J. Am. Chem. Soc.* **127** 12508–9

Zhang X Q, Lam R, Xu X, Chow E K, Kim H J and Ho D 2011 Multimodal nanodiamond drug delivery carriers for selective targeting, imaging, and enhanced chemotherapeutic efficacy *Adv. Mater.* **23** 4770–5

Zhang Y, Zhu Y, Kim G, Wang C, Zhu R, Lu X, Chang H C and Wang Y 2023 Chiral graphene quantum dots enhanced drug loading into exosomes *ACS Nano* **17** 10191–205

Zhao L, Li H and Tan L 2017 A novel fullerene-based drug delivery system delivering doxorubicin for potential lung cancer therapy *J. Nanosci. Nanotechnol.* **17** 5147–54

Zhao D, Zhang R, Liu X, Huang X, Xiao X and Yuan L 2021 One-step synthesis of blue–green luminescent carbon dots by a low-temperature rapid method and their high-performance antibacterial effect and bacterial imaging *Nanotechnology* **32** 155101–1

Zhu C, Du D and Lin Y 2015 Graphene and graphene-like 2D materials for optical biosensing and bioimaging: a review *2D Mater.* **2** 032004

Zou L, Wang S S and Qiu J 2020 Preparation and properties of a glucose biosensor based on an ionic liquid-functionalized graphene/carbon nanotube composite *New Carbon Mater.* **35** 12–9

IOP Publishing

Sustainable Carbon Nanomaterials and their Applications

Rafik Naccache and Adedapo O. Adeola

Chapter 7

Detection of chemical contaminants using eco-friendly carbon nanomaterials

Sifiso A Nsibande and Patricia B C Forbes

Environmental contaminants continue to be a global challenge and require continuous monitoring in order to better understand and mitigate the potential risks they pose to the ecosystem and to human health. The challenge for researchers, therefore, is to develop innovative methods for the sampling, preconcentration, and detection of these contaminants at trace levels. Advances have been made in the development and application of carbon nanomaterials (CNMs) to tackle this challenge. Furthermore, the development of green and eco-friendly synthesis approaches ensures that the production of CNMs has a minimal environmental footprint. This opens the possibility of using these materials to manufacture sampling devices and sensors in a more sustainable manner. The growing interest in using CNMs can be attributed to their excellent properties, which can be easily tuned through various modification strategies. Thus, they have found applications in various fields, including the development of different kinds of sensors (optical sensors, biosensors, and chemical sensors) for the detection of environmental contaminants. CNMs have also been used to sample and preconcentrate environmental contaminants before subjecting them to chromatographic analysis. This chapter highlights the progress that has been made in this regard.

7.1 Introduction

Environmental pollution continues to be a global challenge due to its associated negative effects and potential threats it poses to the ecosystem. Various chemical contaminants, such as heavy metals, herbicides, pesticides, pharmaceuticals, persistent organic pollutants, toxic gases, and many emerging contaminants, are continuously being released into the environment through different anthropogenic activities, thus warranting research into strategies for their sampling, removal, and sensing. The increased research focus on the sampling, detection, removal, and

doi:10.1088/978-0-7503-6325-9ch7   7-1   © IOP Publishing Ltd 2025. All rights, remediation of these contaminants in environmental compartments is encouraging, as it brings general awareness of the problem to the forefront. This, in turn, should lead to better environmental stewardship by regulators, industry players, and communities in general. To this end, eco-friendly CNMs have received significant research attention and effort focused on the design of sampling and detection approaches for various environmental contaminants [1]. The growing interest in the utilization of CNMs is unsurprising, given their well-known fascinating physical and chemical properties including inertness, large surface area, thermal stability, and high conductivity [2]. Moreover, the chemistry of these materials is very versatile, allowing them to be tailored through surface and structural modifications to yield various forms and composite materials with improved and desired properties.

Advancements in the development of simple eco-friendly or green synthesis approaches [3, 4] for producing these nanomaterials make them easily accessible at relatively low cost, which further contributes to their wide adoption. These green approaches seek to avoid the use of toxic chemicals and high-energy input methods in the production of CNMs. Moreover, these methods are just as effective and sometimes better than traditional approaches [5] and can yield the desired product at high quality and in high yields. Eco-friendly synthesis approaches are not only good for the environment, but they also allow for the cost-effective mass production of these materials. This potentially opens doors for the industrial-scale production of future-oriented sampling and sensing devices to address environmental contamination challenges in a much more sustainable manner.

Eco-friendly CNMs have found a number of applications in the detection of environmental contaminants (figure 7.1). In this chapter, therefore, we look at their application in the development of various analytical sensors. These sensors are grouped according to their detection method into sub-groups, namely: (i) colorimetric

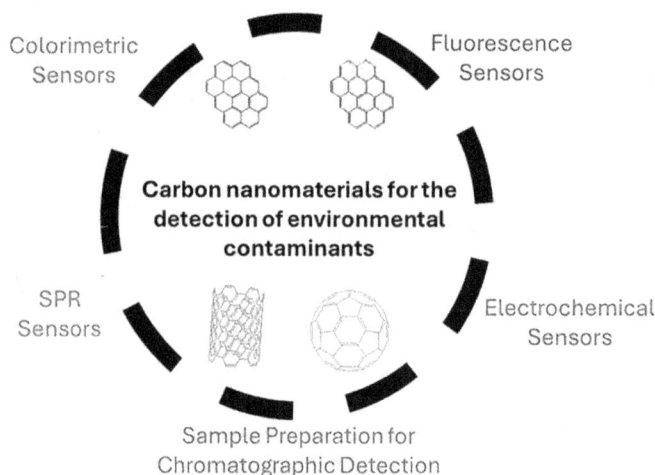

Figure 7.1. Schematic illustration of the different environmental applications of CNMs in detecting environmental contaminants.

sensors, (ii) fluorescence sensors, (iii) surface plasmon resonance (SPR) sensors, and (iv) electrochemical sensors. For the purposes of the current discussion, biosensors fall into each of these groups. Finally, this chapter highlights some of their applications in the sampling and preconcentration of environmental analytes prior to chromatographic detection.

7.2 Sensors employing carbon nanomaterials

There has been growing research interest in the use of CNMs to develop sensors for environmental contaminants. This is due to the potential of these sensors to be cost-effective, particularly for large-scale continuous environmental monitoring campaigns, for which a strong case has been made. The encouraging features of these sensors are their potentially low detection limits (high sensitivity) and their selectivity, which can be achieved through various strategies, including functionalisation with appropriate receptors [6]. In this regard, CNMs have made a significant contribution. The following subsections, therefore, focus on the general construction/synthesis routes and detection principles of the different sensors, as well as some examples of their application in the detection of environmental contaminants.

7.3 Electrochemical sensors

Electrochemical sensors are based on electrochemical reactions or processes that occur at the electrode–solution interface. The general construction can be based on either a two-electrode system (a working electrode and a counter electrode) or a three-electrode system (working, counter, and reference electrodes). Depending on the type of measurement being conducted, electrochemical sensors can be divided into different types [7], as summarized in table 7.1.

Table 7.1. Types of electrochemical sensors. Reproduced from [7], with permission from Springer Nature.

Type of sensor	Summary of its operational principle
Voltammetric (amperometry)	Measures current as a function of applied potential. Potential can be applied continuously or stepwise; hence, voltammetric sensors can be based on cyclic voltammetry (CV), differential pulse voltammetry (DPV), linear sweep voltammetry (LSV), and hydrodynamic voltammetry.
Potentiometric	Measures the potential difference between the reference and the indicator electrode in the absence of current.
Coulometric	Measures and relates the electricity consumed during the reduction/oxidation of the analyte at the electrode interface directly to its mass and therefore its concentration.
Conductometric	Measures the solution's ability to conduct electricity.
Impedimetric	Measures resistance, capacitance, and inductance.

It is perhaps not surprising that CNMs have been extensively used in the design of electrochemical sensors due to their inherently excellent electrical properties. In most cases, the CNMs are used to modify the surface of the working electrode to improve its effective surface area and electron transfer rate, or to serve as catalysts for an electrochemical reaction or process that enhances the detection of the target analyte [8]. CNMs are excellent electrode modifiers due to their large surface area, excellent conductivity, and high adsorption capacity. Furthermore, the ability of carbon to form different allotropes means that novel materials can be tailor-made with unique surface properties for electrochemical sensing applications.

Different methods of electrode modification have been explored, including drop casting and dip coating. A study by Lochab and coworkers [9] reviewed the various applications of CNMs for the electrochemical sensing of heavy metal ions in environmental samples. This report noted that despite the excellent performance characteristics of these sensors, they are still not widely adopted for industrial (commercial) applications due to the potential toxicity of these nanomaterials to humans in the long term.

Electrochemical sensors based on CNMs have also been used for the environmental detection of various other contaminants, a few examples of which are shown in table 7.2. Due to the simplicity of electrochemical cells, there has been a move to develop portable electrochemical sensors based on CNMs, which could be used for the environmental monitoring of contaminants [10]. The progress that has been made in this regard is encouraging, particularly in the quest for the development of field-based monitoring devices. However, challenges such as sensitivity, stability, and finding reproducible protocols for the preparation of the electrodes for such sensors remain. Nanomaterials, particularly CNMs, can play a significant role in improving the analytical performance of these portable devices.

7.4 Fluorescence sensing

Carbon dots (CDs) are a fascinating class of CNMs that have found a number of applications in environmental sensing. According to Yang and coworkers [24], CDs can be subdivided into three further subclasses, namely graphene quantum dots (GQDs), carbon nanodots (CNDs), and polymer dots (PDs). These materials consist of sp^2/sp^3 carbon and oxygen groups which can be doped with different heteroatoms such as nitrogen, sulfur, or phosphorus. These are zero-dimensional materials with sizes typically less than 10 nm.

CDs have unique and fascinating optical properties, particularly their photo-luminescent (PL) properties, which have been exploited extensively in designing fluorescence sensors for target environmental pollutants. The origin and mechanism of their PL property can be attributed to quantum confinement effects, surface/edge states, and molecular/core carbon states [24]. CDs are relatively new members of the carbon family, discovered around 2004 as 'impurities' from the purification of carbon nanotubes (CNTs) [25]. They can be synthesized either via the chemical or physical breakdown of macroscopic graphite-like precursors, named 'top-down synthesis,' or through the treatment of simple molecular precursors, known as

Table 7.2. Examples of studies that employed CNM-based sensors for the detection of environmental contaminants.

Material	Type of sensor/detection mechanism	Analyte	Matrix	LOD	Linear range	References
Graphene oxide/polystyrene sulfonate optical films	Colorimetry	NO_2	—	1.0 ppm	—	[11]
Heteroatom-doped graphene	Colorimetry	Pesticides	—	5–500 μM		[12]
Graphene/Fe_3O_4–Au NPs	Colorimetry	Pb^{2+} ions	Tap water	0.63 ng ml^{-1}	1–300 ng ml^{-1}	[13]
L-cysteine functionalized graphene oxide nanosheet	Colorimetry	Hg^{2+}	Water	7.6 μg l^{-1}	0–200 μg ml^{-1}	[14]
Carbon nanotubes (CNTs)	SPR	2,4-Dichlorophenoxyacetic acid and chlorpyrifos		1.38×10^{-2} ppm and 2.0×10^{-3} ppm		[15]
Graphene quantum dots (GQDs)	Florescence	Pyrene	Water	0.242×10^{-6} M	2–10×10^{-6} M	[16]
Carbon dot nano hybrids	Fluorescence	Hg^{2+} and Pb^{2+}	Water	0.22 nM	1–1000 nM	[17]
Carbon quantum dots (CQDs)	Fluorescence	Cr^{6+}	Aqueous solutions	3.62 ppm	0–500 ppm	[18]
N-dots/molecularly imprinted polymer (MIP)	Fluorescence	Nitroaniline	Lake and groundwater	1.65 nM	5 nM–1 μM	[19]
GQDs@MIP	Fluorescence	Methamphetamine	Water	12 nM	5–50 μM	[20]
N-graphene oxide quantum dots (GOQDs)–Mo2C-MIP	Electrochemical	Indometacin	Water	0.9508×10^{-15} M	10^{-15} to 10^{-5} M	[21]
MIP/C3N4 nanotubes (NTs)@GQDs/glassy carbon electrode (GCE)	Electrochemical (CV)	Chlorpyrifos	Wastewater	0.20 pM	10 pM–1.0 nM	[22]
MIP/Au nanoparticles (NPs)@nitrogen-doped carbon dots (NCDs)@Ag NPs	Electrochemical (DPV)	Metobromuron	Wastewater	0.2 pM	1.0 pM–2.0 nM	[23]

'bottom-up' synthesis. Eco-friendly and sustainable synthesis approaches to produce CDs have also been widely adopted, where biodegradable resources such as biowaste and biomass have been used [26]. Most of the CDs produced via these eco-friendly resources have demonstrated excellent optical properties and could be used in sensing applications without any post-synthesis treatment [27].

Since the discovery of CDs, there has been exponential research interest in their synthesis and application in the detection of environmental contaminants. Sensing is typically based on either the chemical or physical interaction of the analyte with the CDs, thereby altering their surface states, leading to either enhancement or quenching of the PL signal. The most commonly reported sensing mechanisms involve complex formation, the inner filter effect (IFE), Förster resonance energy transfer (FRET), and photoinduced electron transfer (PET). The selectivity of CD-based sensors can be achieved through various strategies, including conjugating them to analyte-specific recognition elements such as molecularly imprinted polymers (MIPs) [28], biological elements (enzymes, antibodies, or DNA aptamers) [29], and supramolecular host molecules [30].

A number of excellent reviews in the literature have been published on the application of CDs for fluorescence sensing of contaminants such as toxic heavy metals [31], polycyclic aromatic hydrocarbons (PAHs) [32], pharmaceutical drugs [33], pesticides [29, 34], and various other environmental contaminants [35].

7.5 Colorimetric sensing

Colorimetric sensing methods can be attractive due to their simplicity, cost-effectiveness, and ease of detection, often through naked-eye observation, thereby reducing reliance on sophisticated instrumentation. The simplicity of colorimetric sensors compared to electrochemical and optical techniques makes them worth exploring. Indeed, various forms of CNMs, their derivatives, and/or hybrids thereof have been explored in the fabrication of colorimetric sensors for environmental applications [36]. One property of CNMs that makes them attractive in designing colorimetric sensors is their typical broad absorption in the visible region and strong polarization-dependent effects. They also have enzyme-mimicking activity (and are thus known as nanozymes), where they facilitate or participate in the catalysis of chemical reactions that are accompanied by a color change. The integration of colorimetric sensors into smartphone cameras with red, green, and blue (RGB) color identification holds great promise in developing portable, low-cost platforms. This has been demonstrated in a study by Zhou and coworkers [37] who developed a cellphone-integrated multichannel sensor as a universal platform for the detection of various contaminants in water. The development of these sensing platforms is a move in the right direction, as they can be widely accessible and thus facilitate citizen science. Their analytical performance can also be further enhanced by the incorporation of CNMs. Colorimetric sensors have also found application in gas sensors. For example, Chi et al [11] used graphene oxide/polystyrene sulfonate (GO/PSS) to develop an optical film which they used for the detection of NO_2, which is a colorless, flammable, and dangerous gas.

7.6 Surface plasmon resonance sensors

SPR optical sensors are attractive as probes for environmental contaminants. In general, the construction of SPR probes consists of a metal dielectric surface (usually gold or silver) to which plane-polarized light is directed and reflected via a prism. The incident beam causes oscillation of the electron density on this metal surface (surface plasmons), which causes attenuation of the reflected beam. The attenuation is further shifted when a monolayer of analyte is present on the surface of the metal. Thus, this spectrometric technique allows for the development of label-free, rapid, and sensitive sensors based on the analytes' binding specificity, association and dissociation kinetics, binding affinity, etc [38].

CNMs can be deposited on the gold or silver surface as thin layer coatings to enhance the electric field intensity on the metal surface and therefore improve the sensitivity of SPR sensors. The optical and electrical properties of CNMs, as well as their large surface area and electron mobility, make them ideal for both signal enhancement and analyte adsorption, which improve detection. As such, SPR has been used in the development of gas sensors for gases such as NO_2, NH_3, and CO_2 [39].

7.7 Carbon nanomaterials in environmental sample preparation for chromatographic detection

Chemical contaminants in environmental compartments typically occur at very low concentrations due to dilution and mixing. This presents a serious analytical challenge, which can be solved by either designing ultrasensitive analytical methods and instrumentation or by developing appropriate sampling and preconcentration protocols that can be employed prior to analysis.

Nanomaterials have played a significant role in the development of sampling and preconcentration methods for environmental pollutants. CNMs such as nanodiamonds, fullerenes, CNTs, graphene, carbon nanofibers, carbon nanocones, graphene wool, and nano-horns, as well as hybrids thereof, have been explored as adsorbents for sampling environmental pollutants in aqueous and gaseous systems [40–42]. Analyte adsorption on these materials can take place through either physisorption or chemisorption.

A significant research effort has focused on the use of CNMs for the removal or remediation (degradation) of contaminants from environmental samples, particularly from aqueous systems [43–45]. Thus, this will not be the focus of this section. Instead, the focus will be on the application of these materials in sampling and preconcentration of target analytes prior to detection, often involving chromatographic separation.

Various forms and modifications of graphene and CNTs can be used in the solid-phase extraction (SPE) of organic contaminants from environmental water samples prior to chromatographic analysis. A review by Azzouz *et al* [46] described different studies in which graphene and CNTs had been used for the SPE of compounds, including PAHs, pesticides, endocrine-disrupting chemicals (EDCs), and volatile organic compounds (VOCs) with satisfactory recoveries and acceptable detection limits. Examples of such studies are presented in table 7.3.

Table 7.3. Examples of studies on the application of CNMs for the extraction and preconcentration of environmental contaminants. Reprinted from [46], Copyright (2018), with permission from Elsevier.

Carbon nanomaterial or composite	Analyte	Matrix	Extraction method	LOD (ng l^{-1})	Recovery (%)	Detection method	References
Graphene	PAHs	Water	Solid-phase extraction (SPE)	0.84–3	72.8–6.2	GC–MS	[47]
Graphene	Chlorophenols	Tap and river water	SPE	100–400	77.2–6.6	HPLC–UV	[48]
Graphene aerogel	Endocrine-disrupting chemicals (EDCs), polychlorinated biphenyls (PCBs)	Water	SPE	10–15	76.3–112.5	GC–MS	[49]
Graphene nanoplatelets	Phthalate esters	Tap water and drinking water	SPE	90–330	87.7–100.9	HPLC DAD	[50]
Graphene oxide	Chlorophenoxy acid herbicides	Water	SPE	300–500	75–77	Capillary electrophoresis	[51]
Graphene oxide framework	Phenylurea herbicides	Water	SPE	10–20	–	HPLC	[52]
Reduced graphene oxide (rGO)-silica	Fluoroquinolones	Tap and river water	SPE	2	72–118	LC–FLD	[53]
Multiwalled carbon nanotubes (MWCNTs)	Carbamate pesticides	Surface water	SPE	10–50	92.22–103.9	LC–MS	[54]
Packed MWCNTs	β-blockers and nonsteroidal anti-inflammatory drugs (NSAIDs)	River and wastewater	SPE	9.0–121	68–107	LC–MS/MS	[55]

Material	Analyte	Extraction	Matrix	Range	Recovery (%)	Detection	Ref.
MWCNT disks	Atrazine and simazine	SPE	Water	2.5–5.0	—	GC–MS	[56]
Planar GO-based magnetic ionic liquid nanomaterial	Chlorophenols	Magnetic solid-phase extraction (MSPE)	Tap water, river water, well water	0.2–8.7	85.3–99.3	LC–MS/MS	[57]
Magnetized graphene (m-graphene)/carbon nanofiber (CNF)	PAHs	MSPE	Water	4.0–30	95.5–99.9	GC–FID	[58]
Fe_3O_4@SiO_2-Graphene	Phthalate esters	MSPE	Water	70–100	87.2–109.0	HPLC–UV	[59]
Graphene-C_3N_4/Fe_3O_4	PAHs	MSPE	Water	50–100	80.0–99.8	HPLC–UV	[60]
Fe_3O_4@SiO_2@GO composite	PAHs	MSPE	Water	5.0–100	71.7–106.7	GC–FID	[61]
Magnetic ionic-liquid-modified MWCNTs	Aryloxyphenoxy-propionate herbicides and their metabolites	MSPE	Water	2.8–43.2	66.1–89.6	HPLC DAD	[62]
Fe_3O_4–CNTs	Phthalate acid esters	MSPE	Water	4.9–38	64.6–125.6	GC–MS	[63]
Fe_3O_4–MWCNTs	Phthalate esters	MSPE	Water	9.0–32	86.6–100.2	GC–MS/MS	[64]
Fe_3O_4–MWCNTs	Alkylbenzene sulfonates	MSPE	Water	13–21	87.3–106.3	HPLC–UV	[65]

PleGas chromatography mass spectrometry (GC-MS); High-pressure liquid chromatography with ultra-violent detection (HPLC-UV); high-pressure liquid chromatography with diode-ultra-violent detection (HPLC-DAD); liquid chromatography with fluorescence detection (LC-FLD); liquid chromatography mass spectrometry (LC-MS); liquid chromatography tandem mass spectrometry (LC-MS/MS); gas chromatography with flame ionization detection (GC-FID).

Porous carbon nanomaterials (PCNMs) are another class of CNMs that have been explored for use in sample preparation and the preconcentration of environmental contaminants prior to chromatographic detection. PCNMs consist of three main subclasses, namely microporous (<2 nm pore size), mesoporous (2–50 nm pore size), and macroporous (>50 nm pore size) PCNMs. The pore sizes of these materials can be modified through various synthesis strategies, including: (i) the hard template method, (ii) the soft template method, (iii) the self-template method, (iv) the activated method, and (v) the calcination method. PCNMs have been used in SPE, solid-phase microextraction (SPME), dispersive solid-phase extraction (DSPE), and magnetic solid-phase extraction (MSPE) of various organic contaminants from environmental samples [66].

CNMs have also been incorporated into SPME devices. For example, a study by Cardeal and coworkers [67] used chemical vapor deposition (CVD) techniques to coat SPME fibers with a CNM (graphene), thereby developing a device that was used to extract and preconcentrate pesticides from water samples. Using this graphene-coated SPME device, they could extract and detect 24 different pesticides from water using GC–MS and obtained detection limits ranging from 0.0002 to 1.1309 μg l^{-1}. The excellent performance of the device was attributed to the adsorption capacity of the CNM coating as well as faster equilibration during extraction. In another study, Ouyang and coworkers [68] used CDs to modify their SPME fibers, producing a fluorescent SPME device that could be used both for the preconcentration of 2-nitroaniline before GC–MS detection and fluorescence detection/sensing thereof. Using this device, the authors harnessed the power of the two detection methods and were able to quantify 2-nitroaniline down to 0.011 μg l^{-1} in water and urine samples.

CNMs generally have a great affinity for most organic contaminants, as has been shown in remediation studies. However, the lack of selectivity can be a challenge, particularly for complex matrices. To overcome this, attempts have been made to functionalize the CNMs with MIPs [69]. While this strategy can enhance the selectivity and specificity of the CNMs in sample preparation, some challenges remain, including the leaching of monomers, which can complicate the chromatograms, as well as the extra cost associated with the MIP modification step. Another limitation on the application of CNMs in sampling and sample preparation is the practicality of their handling. To overcome this challenge, researchers have been incorporating them into various supports, including membranes [70], which can then be used for the SPE of the target analytes. The handling challenge can also be overcome by incorporating the CNMs into easy-to-handle substrates. This was demonstrated by Forbes and coworkers [42], who used quartz wool as a support for graphene to develop a novel sampler for volatile and semi-volatile organic compounds. The resulting graphene wool, produced via CVD [71], was easy to handle and showed excellent sampling efficiency, which compared well with other materials employed in air sampling devices, such as polydimethylsiloxane (PDMS).

7.8 Conclusions and prospects

Due to the versatility of CNMs, they have attracted a lot of research interest and applications in solving environmental challenges. Because of their attractive and tunable properties, they continue to be explored for the sampling and detection of environmental chemical contaminants in different environmental compartments, such as soil, water, sediments, and air. The future-centric approach of using green synthesis methods is helping to ensure sustainable, large-scale synthesis of these materials for industrial applications. The different sensing strategies have been shown to have excellent analytical performance (sensitivity and selectivity) and offer advantages over traditional chromatography-based approaches, particularly regarding cost. Despite this, several hurdles still need to be overcome to achieve universal industrial adoption of these sensors.

Future studies, therefore, should consider the following prospects:

- There is still potential for the detection limits of carbon-based sensors, particularly optical sensors, to be further pushed even lower by introducing some form of sample preparation and/or preconcentration, such as SPE, prior to detection. This is notwithstanding, of course, the fact that most of these sensors are intended to avoid precisely this step, but adding it would enhance detection limits.

- The practical operational convenience of CNM-based sensors remains a challenge that scientists need to confront. This could be addressed through miniaturization approaches where simple and portable prototype devices are developed with the potential for in-field analysis. Currently, most of the excellent sensors still rely on sophisticated laboratory bench-top instrumentation.

- Colorimetric CNM-based sensors using smartphone detection have the potential for use in citizen science, which would allow for widespread collection of data related to environmental contamination and early warning of pollution events.

- Mass production of the sensors should be a future consideration. The use of eco-friendly CNMs from natural and reusable resources, such as biowaste, is a move in the right direction to achieve this future objective, as it will ensure that mass production is cost-effective and sustainable.

- More elaborate CNMs and their composites will be developed and used in sampling devices and in sample preparation for established chromatography-based detection methods. Eco-friendly CNMs can therefore play a significant role in reducing the costs of analysis of these methods.

References

[1] Poonia K *et al* 2024 Sustainability, performance, and production perspectives of waste-derived functional carbon nanomaterials towards a sustainable environment: a review *Chemosphere* **352** 141419

[2] Gopinath K P, Vo D-V N, Gnana Prakash D, Adithya Joseph A, Viswanathan S and Arun J 2021 Environmental applications of carbon-based materials: a review *Environ. Chem. Lett.* **19** 557–82

[3] Verma S K, Das A K, Gantait S, Panwar Y, Kumar V and Brestic M 2022 Green synthesis of carbon-based nanomaterials and their applications in various sectors: a topical review *Carbon Lett.* **32** 365–93

[4] Gupta D, Boora A, Thakur A and Gupta T K 2023 Green and sustainable synthesis of nanomaterials: recent advancements and limitations *Environ. Res.* **231** 116316

[5] Huston M, DeBella M, DiBella M and Gupta A 2021 Green synthesis of nanomaterials *Nanomaterials* **11** 2130

[6] Willner M R and Vikesland P J 2018 Nanomaterial-enabled sensors for environmental contaminants *J. Nanobiotechnol.* **16** 95

[7] Saputra H A 2023 Electrochemical sensors: basic principles, engineering, and state of the art *Monatsh. Chem.—Chem. Monthly* **154** 1083–100

[8] Sahragard A, Varanusupakul P and Miró M 2023 Nanomaterial decorated electrodes in flow-through electrochemical sensing of environmental pollutants: a critical review *Trends Environ. Anal. Chem.* **39** e00208

[9] Lochab A, Sharma R, Kumar S and Saxena R 2021 Recent advances in carbon based nanomaterials as electrochemical sensor for toxic metal ions in environmental applications *Mater. Today Proc.* **45** 3741–53

[10] He Q, Wang B, Liang J, Liu J, Liang B, Li G, Long Y, Zhang G and Liu H 2023 Research on the construction of portable electrochemical sensors for environmental compounds quality monitoring *Mater. Today Adv.* **17** 100340

[11] Chi H, Xu Z, Duan X, Yang J, Wang F and Li Z 2019 High-performance colorimetric room-temperature NO2 sensing using spin-coated graphene/polyelectrolyte reflecting film *ACS Appl. Mater. Interfaces* **11** 32390–7

[12] Zhu Y, Wu J, Han L, Wang X, Li W, Guo H and Wei H 2020 Nanozyme sensor arrays based on heteroatom-doped graphene for detecting pesticides *Anal. Chem.* **92** 7444–52

[13] Tao Z, Zhou Y, Duan N and Wang Z 2020 A colorimetric aptamer sensor based on the enhanced peroxidase activity of functionalized graphene/Fe_3O_4-AuNPs for detection of lead (II) ions *Catalysts* **10** 600

[14] Tian H, Liu J, Guo J, Cao L and He J 2022 L-Cysteine functionalized graphene oxide nanoarchitectonics: a metal-free Hg^{2+} nanosensor with peroxidase-like activity boosted by competitive adsorption *Talanta* **242** 123320

[15] Wang Y, Cui Z, Zhang X, Zhang X, Zhu Y, Chen S and Hu H 2020 Excitation of surface plasmon resonance on multiwalled carbon nanotube metasurfaces for pesticide sensors *ACS Appl. Mater. Interfaces* **12** 52082–8

[16] Nsibande S A and Forbes P B C 2020 Development of a turn-on graphene quantum dot-based fluorescent probe for sensing of pyrene in water *RSC Adv.* **10** 12119–28

[17] Liu Z, Jin W, Wang F, Li T, Nie J, Xiao W, Zhang Q and Zhang Y 2019 Ratiometric fluorescent sensing of Pb^{2+} and Hg^{2+} with two types of carbon dot nanohybrids synthesized from the same biomass *Sensors Actuators* B **296** 126698

[18] Baragau I-A, Power N P, Morgan D J, Lobo R A, Roberts C S, Titirici M-M, Middelkoop V, Diaz A, Dunn S and Kellici S 2021 Efficient continuous hydrothermal flow synthesis of carbon quantum dots from a targeted biomass precursor for on–off metal ions nanosensing *ACS Sustain. Chem. Eng.* **9** 2559–69

[19] Nie Y, Liu Y, Su X and Ma Q 2019 Nitrogen-rich quantum dots-based fluorescence molecularly imprinted paper strip for p-nitroaniline detection *Microchem. J.* **148** 162–8

[20] Masteri-Farahani M, Mashhadi-Ramezani S and Mosleh N 2020 Molecularly imprinted polymer containing fluorescent graphene quantum dots as a new fluorescent nanosensor for detection of methamphetamine *Spectrochim. Acta, Part* A **229** 118021

[21] Lu H, Cui H, Duan D, Li L and Ding Y 2021 A novel molecularly imprinted electrochemical sensor based on a nitrogen-doped graphene oxide quantum dot and molybdenum carbide nanocomposite for indometacin determination *Analyst* **146** 7178–86

[22] Yola M L and Atar N 2017 A highly efficient nanomaterial with molecular imprinting polymer: carbon nitride nanotubes decorated with graphene quantum dots for sensitive electrochemical determination of chlorpyrifos *J. Electrochem. Soc.* **164** B223–9

[23] Kıran T R, Yola M L and Atar N 2019 Electrochemical sensor based on Au@nitrogen-doped carbon quantum dots@Ag core-shell composite including molecular imprinted polymer for metobromuron recognition *J. Electrochem. Soc.* **166** H691–7

[24] Zhu S, Song Y, Zhao X, Shao J, Zhang J and Yang B 2015 The photoluminescence mechanism in carbon dots (graphene quantum dots, carbon nanodots, and polymer dots): current state and future perspective *Nano Res.* **8** 355–81

[25] Xu X, Ray R, Gu Y, Ploehn H J, Gearheart L, Raker K and Scrivens W A 2004 Electrophoretic analysis and purification of fluorescent single-walled carbon nanotube fragments *J. Am. Chem. Soc.* **126** 12736–7

[26] Dhariwal J, Rao G K and Vaya D 2024 Recent advancements towards the green synthesis of carbon quantum dots as an innovative and eco-friendly solution for metal ion sensing and monitoring *RSC Sustain.* **2** 11–36

[27] Ullal N, Muthamma K and Sunil D 2022 Carbon dots from eco-friendly precursors for optical sensing application: an up-to-date review *Chem. Papers* **76** 6097–127

[28] Sobiech M, Luliński P, Wieczorek P P and Marć M 2021 Quantum and carbon dots conjugated molecularly imprinted polymers as advanced nanomaterials for selective recognition of analytes in environmental, food and biomedical applications *TrAC, Trends Anal. Chem.* **142** 116306

[29] Su D, Li H, Yan X, Lin Y and Lu G 2021 Biosensors based on fluorescence carbon nanomaterials for detection of pesticides *TrAC, Trends Anal. Chem.* **134** 116126

[30] Yan F, Hou Y, Yi C, Wang Y, Xu M and Xu J 2022 Carbon dots modified/prepared by supramolecular host molecules and their potential applications: a review *Anal. Chim. Acta* **1232** 340475

[31] Devi P, Rajput P, Thakur A, Kim K-H and Kumar P 2019 Recent advances in carbon quantum dot-based sensing of heavy metals in water *TrAC, Trends Anal. Chem.* **114** 171–95

[32] Nsibande S A, Montaseri H and Forbes P B C 2019 Advances in the application of nanomaterial-based sensors for detection of polycyclic aromatic hydrocarbons in aquatic systems *TrAC, Trends Anal. Chem.* **115** 52–69

[33] Rajendran S, UshaVipinachandran V, Badagoppam Haroon K H, Ashokan I and Bhunia S K 2022 A comprehensive review on multi-colored emissive carbon dots as fluorescent probes for the detection of pharmaceutical drugs in water *Anal. Methods* **14** 4263–91

[34] Nsibande S A and Forbes P B C 2016 Fluorescence detection of pesticides using quantum dot materials—a review *Anal. Chim. Acta* **945** 9–22

[35] Ajith M P, Pardhiya S and Rajamani P 2022 Carbon dots: an excellent fluorescent probe for contaminant sensing and remediation *Small* **18** 2105579

[36] Wu Y, Feng J, Hu G, Zhang E and Yu H-H 2023 Colorimetric sensors for chemical and biological sensing applications *Sensors* **23** 2749

[37] Xing Y, Xue B, Lin Y, Wu X, Fang F, Qi P, Guo J and Zhou X 2022 A cellphone-based colorimetric multi-channel sensor for water environmental monitoring *Front. Environ. Sci. Eng.* **16** 155

[38] Zhang P, Chen Y-P, Wang W, Shen Y and Guo J-S 2016 Surface plasmon resonance for water pollutant detection and water process analysis *TrAC, Trends Anal. Chem.* **85** 153–65

[39] Kumar V, Raghuwanshi S K and Kumar S 2022 Recent advances in carbon nanomaterials based SPR sensor for biomolecules and gas detection—a review *IEEE Sens. J.* **22** 15661–72

[40] Hussain C M 2014 Carbon nanomaterials as adsorbents for environmental analysis *Nanomaterials for Environmental Protection* 217–36

[41] Zhang B-T, Zheng X, Li H-F and Lin J-M 2013 Application of carbon-based nanomaterials in sample preparation: a review *Anal. Chim. Acta* **784** 1–17

[42] Geldenhuys G-L, Mason Y, Dragan G C, Zimmermann R and Forbes P 2021 Novel graphene wool gas adsorbent for volatile and semivolatile organic compounds *ACS Omega* **6** 24765–76

[43] Adeola A O, Duarte M P and Naccache R 2023 Microwave-assisted synthesis of carbon-based nanomaterials from biobased resources for water treatment applications: emerging trends and prospects *Front. Carbon* **2**

[44] Manimegalai S, Vickram S, Deena S R, Rohini K, Thanigaivel S, Manikandan S, Subbaiya R, Karmegam N, Kim W and Govarthanan M 2023 Carbon-based nanomaterial intervention and efficient removal of various contaminants from effluents—a review *Chemosphere* **312** 137319

[45] Adeola A O and Forbes P B C 2024 Two-dimensional carbon-based materials for sorption of selected aromatic compounds in water *Carbon Nanomaterials and their Composites as Adsorbents* ed J Tharini and S Thomas (Cham: Springer International Publishing) pp 247–60

[46] Azzouz A, Kailasa S K, Lee S S, J. Rascón A, Ballesteros E, Zhang M and Kim K-H 2018 Review of nanomaterials as sorbents in solid-phase extraction for environmental samples *TrAC, Trends Anal. Chem.* **108** 347–69

[47] Wang Z, Han Q, Xia J, Xia L, Ding M and Tang J 2013 Graphene-based solid-phase extraction disk for fast separation and preconcentration of trace polycyclic aromatic hydrocarbons from environmental water samples *J. Sep. Sci.* **36** 1834–42

[48] Liu Q, Shi J, Zeng L, Wang T, Cai Y and Jiang G 2011 Evaluation of graphene as an advantageous adsorbent for solid-phase extraction with chlorophenols as model analytes *J. Chromatogr.* A **1218** 197–204

[49] Han Q, Liang Q, Zhang X, Yang L and Ding M 2016 Graphene aerogel based monolith for effective solid-phase extraction of trace environmental pollutants from water samples *J. Chromatogr.* A **1447** 39–46

[50] Luo X, Zhang F, Ji S, Yang B and Liang X 2014 Graphene nanoplatelets as a highly efficient solid-phase extraction sorbent for determination of phthalate esters in aqueous solution *Talanta* **120** 71–5

[51] Tabani H, Fakhari A R, Shahsavani A, Behbahani M, Salarian M, Bagheri A and Nojavan S 2013 Combination of graphene oxide-based solid phase extraction and electro membrane extraction for the preconcentration of chlorophenoxy acid herbicides in environmental samples *J. Chromatogr.* A **1300** 227–35

[52] Li M, Wang J, Jiao C, Wang C, Wu Q and Wang Z 2016 Graphene oxide framework: an adsorbent for solid phase extraction of phenylurea herbicides from water and celery samples *J. Chromatogr.* A **1469** 17–24

[53] Speltini A, Sturini M, Maraschi F, Consoli L, Zeffiro A and Profumo A 2015 Graphene-derivatized silica as an efficient solid-phase extraction sorbent for pre-concentration of fluoroquinolones from water followed by liquid-chromatography fluorescence detection *J. Chromatogr. A* **1379** 9–15

[54] Latrous El Atrache L, Hachani M and Kefi B B 2016 Carbon nanotubes as solid-phase extraction sorbents for the extraction of carbamate insecticides from environmental waters *Int. J. Environ. Sci. Technol.* **13** 201–8

[55] Dahane S, Gil García M D, Martínez Bueno M J, Uclés Moreno A, Martínez Galera M and Derdour A 2013 Determination of drugs in river and wastewaters using solid-phase extraction by packed multi-walled carbon nanotubes and liquid chromatography–quadrupole-linear ion trap-mass spectrometry *J. Chromatogr. A* **1297** 17–28

[56] Katsumata H, Kojima H, Kaneco S, Suzuki T and Ohta K 2010 Preconcentration of atrazine and simazine with multiwalled carbon nanotubes as solid-phase extraction disk *Microchem. J.* **96** 348–51

[57] Cai M-Q, Su J, Hu J-Q, Wang Q, Dong C-Y, Pan S-D and Jin M-C 2016 Planar graphene oxide-based magnetic ionic liquid nanomaterial for extraction of chlorophenols from environmental water samples coupled with liquid chromatography–tandem mass spectrometry *J. Chromatogr. A* **1459** 38–46

[58] Rezvani-Eivari M, Amiri A, Baghayeri M and Ghaemi F 2016 Magnetized graphene layers synthesized on the carbon nanofibers as novel adsorbent for the extraction of polycyclic aromatic hydrocarbons from environmental water samples *J. Chromatogr. A* **1465** 1–8

[59] Wang W, Ma R, Wu Q, Wang C and Wang Z 2013 Fabrication of magnetic microsphere-confined graphene for the preconcentration of some phthalate esters from environmental water and soybean milk samples followed by their determination by HPLC *Talanta* **109** 133–40

[60] Wang M, Cui S, Yang X and Bi W 2015 Synthesis of g-C3N4/Fe3O4 nanocomposites and application as a new sorbent for solid phase extraction of polycyclic aromatic hydrocarbons in water samples *Talanta* **132** 922–8

[61] Mahpishanian S, Sereshti H and Ahmadvand M 2017 A nanocomposite consisting of silica-coated magnetite and phenyl-functionalized graphene oxide for extraction of polycyclic aromatic hydrocarbon from aqueous matrices *J. Environ. Sci.* **55** 164–73

[62] Luo M, Liu D, Zhao L, Han J, Liang Y, Wang P and Zhou Z 2014 A novel magnetic ionic liquid modified carbon nanotube for the simultaneous determination of aryloxyphenoxypropionate herbicides and their metabolites in water *Anal. Chim. Acta* **852** 88–96

[63] Luo Y-B, Yu Q-W, Yuan B-F and Feng Y-Q 2012 Fast microextraction of phthalate acid esters from beverage, environmental water and perfume samples by magnetic multi-walled carbon nanotubes *Talanta* **90** 123–31

[64] Jiao Y, Fu S, Ding L, Gong Q, Zhu S, Wang L and Li H 2012 Determination of trace leaching phthalate esters in water by magnetic solid phase extraction based on magnetic multi-walled carbon nanotubes followed by GC–MS/MS *Anal. Methods* **4** 2729–34

[65] Chen B, Wang S, Zhang Q and Huang Y 2012 Highly stable magnetic multiwalled carbon nanotube composites for solid-phase extraction of linear alkylbenzene sulfonates in environmental water samples prior to high-performance liquid chromatography analysis *Analyst* **137** 1232–40

[66] Wang Y, Chen J, Ihara H, Guan M and Qiu H 2021 Preparation of porous carbon nanomaterials and their application in sample preparation: a review *TrAC, Trends Anal. Chem.* **143** 116421

[67] Valenzuela E F, de Paula F G F, Teixeira A P C, Menezes H C and Cardeal Z L 2020 A new carbon nanomaterial solid-phase microextraction to pre-concentrate and extract pesticides in environmental water *Talanta* **217** 121011

[68] Liu S, Yang H, Yang S, Huang Y, Xiang Z and Ouyang G 2020 Carbon dots based solid phase microextraction of 2-nitroaniline followed by fluorescence sensing for selective early screening and sensitive gas chromatography-mass spectrometry determination *Anal. Chim. Acta* **1111** 147–54

[69] Singh M, Singh S, Singh S P and Patel S S 2020 Recent advancement of carbon nanomaterials engrained molecular imprinted polymer for environmental matrix *Trends Environ. Anal. Chem.* **27** e00092

[70] Bosco C D, De Cesaris M G, Felli N, Lucci E, Fanali S and Gentili A 2023 Carbon nanomaterial-based membranes in solid-phase extraction *Microchim. Acta* **190** 175

[71] Schoonraad G-L, Madito M J, Manyala N and Forbes P 2020 Synthesis and optimisation of a novel graphene wool material by atmospheric pressure chemical vapour deposition *J. Mater. Sci.* **55** 545–64

IOP Publishing

Sustainable Carbon Nanomaterials and their Applications

Rafik Naccache and Adedapo O. Adeola

Chapter 8

Harnessing carbon-based materials for energy production and conversion

Adedipe Demilade, Dorcas Adenuga,
Pannan I Kyesmen and Kayode Adesina Adegoke

The increasing global demand for clean, sustainable energy, driven by population growth and industrialization, has necessitated the transition from fossil fuels to renewable energy sources. However, the intermittent nature of renewables requires advanced energy conversion and storage technologies to ensure reliability. Carbon-based materials, including graphene, carbon nanotubes (CNTs), and biomass-derived activated carbon, have emerged as pivotal components in this energy transition due to their tunable physicochemical properties, high conductivity, large surface area, and environmental compatibility. This review explores the versatile roles of carbon-based materials in energy production and storage systems, such as fuel cells, photovoltaic cells, supercapacitors, and batteries. It discusses their use in electrocatalytic reactions, oxygen reduction, hydrogen evolution, and CO_2 reduction, as well as their integration in thermoelectric devices and biofuel production. Functionalization and hybridization strategies further enhance their performance, enabling metal-free catalysis and improved energy densities. The environmental sustainability of these materials is also assessed through life cycle analyses, green synthesis methods, and recycling initiatives. Despite significant advancements, challenges remain in catalyst preparation, stability, scalability, and cost. This review identifies key research gaps and highlights opportunities for optimizing carbon-based materials in next-generation energy technologies through interdisciplinary innovation.

8.1 Overview of carbon-based materials: the importance of energy production and storage

The escalating global energy demand, propelled by population expansion and industrial advancement, has increased the necessity for sustainable energy solutions. Fossil fuels, having served as the primary energy source for more than a hundred

doi:10.1088/978-0-7503-6325-9ch8

8-1

years, are not only limited in availability but also contribute significantly to greenhouse gas emissions and environmental degradation [1]. In this context, the shift toward renewable energy sources, including solar, wind, and biomass, is essential. Nevertheless, the intermittent/variable nature of these sources requires the implementation of effective energy storage systems to maintain a consistent energy supply [2, 3]. The progress of energy technologies is strongly dependent on the evolution of materials that exhibit enhanced environmental compatibility, robust performance, and improved scalability. Among these, carbon-based materials have become essential elements in the advancement of next-generation energy technologies, owing to their distinctive physicochemical characteristics, structural tunability, and commitment to environmental sustainability. Substances such as graphene, CNTs, activated carbon (biomass), and their derivatives exhibit significant surface area, remarkable electrical conductivity, chemical stability, and mechanical strength, rendering them appropriate for various applications in energy production and storage systems (figure 8.1) [4–10]. The adaptability of carbon materials arises from their capacity to form diverse allotropes and hybrid structures, enabling meticulous regulation of their morphology and electronic characteristics at the nanoscale. These characteristics render them exceptionally suitable for incorporation into fuel cells, batteries, supercapacitors, and photovoltaic devices.

The escalating worldwide demand for clean, dependable, and sustainable energy has highlighted the pressing necessity for advanced energy production and storage technologies. The shift from fossil fuels to renewable energy sources, including wind and solar, requires the implementation of robust energy storage systems to address intermittency and maintain grid stability [11, 12]. Within this framework, energy storage devices—especially photo/electrochemical systems such as lithium-ion batteries (LIBs) and supercapacitors—are of paramount importance. The characteristics of electrode and electrolyte materials largely determine their performance, longevity, and ecological footprint. Carbon-based materials are the subject of extensive research due to their superior catalytic performance and compatibility with various architectures, which can significantly improve energy density, power output, and cycling stability [13–15].

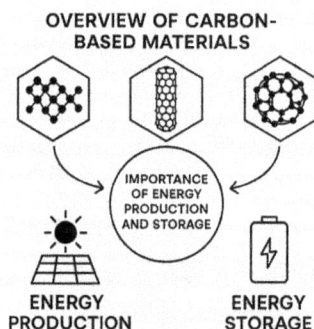

Figure 8.1. Overview of carbon-based materials, showing the importance of energy production and storage.

Simultaneously, carbon-based materials play a significant role in energy production technologies, including electrocatalysis and photoelectrochemical (PEC) cells. For example, composites of doped graphene and CNTs have demonstrated considerable potential in facilitating the oxygen reduction reaction (ORR) and the hydrogen evolution reaction (HER), which are crucial for applications in fuel cells and water-splitting processes [16, 17]. Moreover, the functionalization of carbon materials facilitates the adjustment of their electronic structure and catalytic performance, thereby creating avenues for the development of economical, metal-free catalysts. The incorporation of carbon-based nanomaterials within energy systems enhances device efficiency while simultaneously aligning with the principles of environmental sustainability and the abundance of materials.

As energy technologies advance to address the challenges of a decarbonized future, carbon-based materials will remain essential. Current investigations are centered on scalable synthesis techniques, structural refinement, and hybridization approaches to further elevate their efficacy across a range of energy applications. The integration of materials science, catalysis, and nanotechnology in this field highlights the multifaceted approach required to enhance energy storage and conversion systems via carbon-based advancements.

8.1.1 Carbon-based materials: definition and classification

Carbon-based materials encompass a diverse array of substances primarily constituted of carbon atoms organized in a multitude of allotropic and composite configurations. These encompass conventional forms such as graphite and activated carbon derived from biomass, in addition to nanostructured materials including CNTs, graphene, carbon nanofibers, etc [8]. The adaptability of carbon is evident in its capacity to create sp, sp^2, and sp^3 hybridized bonds, facilitating the development of one-dimensional (1D), two-dimensional (2D), and three-dimensional (3D) structures with adjustable characteristics. Graphene, consisting of a single atomic layer of sp^2-bonded carbon atoms organized in a hexagonal lattice, demonstrates remarkable thermal and electrical conductivity, impressive mechanical strength, and an extensive specific surface area [1]. CNTs, which are cylindrical formations made from rolled graphene sheets, exhibit remarkable tensile strength and conductivity, rendering them appropriate for a variety of structural and electronic applications [3]. Activated carbon derived from biomass, noted for its significant porosity and expansive surface area, finds extensive application in the realms of adsorption and electrochemical energy storage [18]. These materials can be meticulously designed at the nanoscale to augment particular functionalities essential for energy systems.

8.1.2 Importance of carbon-based materials in energy production

In the field of energy production, carbon-based materials significantly contribute to the optimization and efficacy of energy conversion devices. Graphene and CNTs have been utilized in the advancement of next-generation photovoltaic devices, enhancing charge carrier mobility and light absorption efficiency [19, 20]. In fuel cells, carbon supports that are doped with heteroatoms like sulfur or nitrogen have demonstrated enhanced catalytic activity for the HER, the ORR, and the CO_2 reduction reaction (CO_2RR) [21, 22]. Furthermore, carbon-based electrodes have been employed in microbial fuel cells and electrochemical water-splitting systems, where their superior conductivity and chemical stability enhance energy conversion efficiencies [23]. Additionally, carbon-based materials play a crucial role in thermo-electric devices that facilitate the conversion of waste heat into biofuels, biodiesels, and/or electricity. Their low thermal conductivity, when coupled with improved electrical conductivity, yields a favorable thermoelectric figure of merit, particularly when integrated with other nanostructures [24]. The flexibility inherent in carbon materials facilitates the precise adjustment of energy bandgaps and surface func-tionalities, aligning them with particular energy conversion needs.

8.1.3 Roles of carbon-based materials in energy storage

The efficacy of renewable energy systems is intricately linked to the performance and capability of energy storage technologies. Carbon-based materials play a pivotal role in the advancement of high-performance batteries and supercapacitors, attributed to their electrochemical stability, conductivity, and extensive surface area. In LIBs, graphite is utilized as the conventional anode material owing to its capacity to intercalate lithium ions while maintaining structural integrity with minimal degra-dation [12]. Investigations are currently exploring graphene and porous carbon architectures as potential substitutes or enhancements for graphite, with the goal of achieving superior energy densities and accelerated charge–discharge rates [25].

In supercapacitors, commonly referred to as electrochemical capacitors, carbon materials play a crucial role owing to their fast charge transport and storage abilities facilitated by electric double-layer capacitance (EDLC) and pseudo-capacitance mechanisms [13]. The hierarchical pore structure of activated carbon facilitates efficient ion diffusion and storage, establishing it as a fundamental material in commercial supercapacitors [13, 26]. Composites derived from carbon, achieved through the amalgamation of metal oxides or conductive polymers with carbon matrices, present improved charge storage capacities and stability [10]. Sodium-ion batteries, lithium–sulfur batteries, and metal–air batteries derive significant advan-tages from carbon-based architectures, which offer mechanical support, improve electron mobility, and stabilize electrochemical interfaces [27]. Recent advancements highlight the crucial significance of carbon materials in overcoming the constraints of existing energy storage technologies, especially in attaining enhanced energy density, prolonged cycle life, and improved safety.

8.1.4 Sustainability and scalability of carbon-based materials

The deployment of materials for energy applications necessitates a profound consideration of sustainability. Carbon-based materials present notable benefits in this area, owing to their capacity for synthesis from renewable biomass, waste materials, and various other plentiful precursors [14]. For example, activated carbon can be obtained from agricultural biomass residues, whereas graphene and CNTs can be produced through methods such as chemical vapor deposition or liquid-phase exfoliation, which are progressively optimized for enhanced scalability and environmental safety [28]. Moreover, the capacity for recycling and the minimal toxicity of carbon materials render them significantly more environmentally friendly compared to numerous metal-based alternatives. Life cycle assessments (LCAs) indicate that carbon-based electrodes frequently yield reduced environmental footprints, especially when derived from sustainable feedstocks [29]. Nevertheless, obstacles persist in the domains of mass production, uniformity in material quality, and economically viable incorporation into commercial energy frameworks.

8.1.5 Scope of this chapter

This chapter provides an in-depth analysis of the role of carbon-based materials in energy generation and storage systems. Section 8.2 delves into biofuel production, exploring the overview of biofuels and their importance, the role of carbon-based catalysts in biofuel synthesis, case studies in biofuel production, and the production of biodiesel from biomass. Section 8.3 discusses catalytic processes in CO_2 conversion, covering the mechanisms of CO_2 conversion, case studies of CO_2 conversion using carbon nanomaterials/composites, photocatalytic CO_2 reduction using carbon-based nanomaterials/composites, and the use of carbon in energy-efficient catalysis. Section 8.4 focuses on green energy production, including the use of various carbon-based materials in solar energy conversion and carbon nanomaterials in hydrogen production (water splitting and biomass gasification). Section 8.5 presents the concluding remarks, addressing the environmental impact and sustainability of carbon-based materials along with challenges, research gaps, and opportunities. Through this book, the reader will be able to take advantage of the central role of carbon-based materials in advancing the global energy transition and fostering a sustainable energy future.

8.2 Biofuel production

8.2.1 Overview of biofuels and their importance

Energy-rich chemicals made from biomass derived from plants and microbes through a range of biochemical, physical, and thermochemical processes make up biofuels [30, 31]. The amount of biofuel produced worldwide rose by 17% between 2009 and 2010, reaching 105 billion liters (28 billion US gallons) in 2010. Biofuels, which mainly consist of ethanol and biodiesel, accounted for 2.7% of all road transportation fuels worldwide [32]. Brazil and the United States are the two largest users and suppliers of bioethanol, accounting for over 87.5% of global output. To produce low-cost bioethanol, the United States has committed a significant portion of its incredibly productive agricultural areas to the production of maize, whereas Brazil uses sugarcane as a raw material [32].

Natural gas, liquefied petroleum gas, hydrogen, unconventional fossil fuels, electricity, Fischer–Tropsch liquids, ethers, alcohols, biodiesels, and more have all been explored as alternative fuels over the years. However, biofuels, which can be made directly from biomass, have been a viable option among these test fuels [32, 33]. Additionally, due to environmental concerns and consequences, the depletion of these fuel supplies has forced the use of renewable biofuels as an energy source, which has become increasingly important in the last 20 years [34]. Interest in promoting biofuels as one of the top renewable energy sources has also been sparked by the release of carbon dioxide from the burning of fossil fuels, the primary contributor to this process. The natural balance of the ecosystem is disturbed when greenhouse gases are released from the transportation sector's combustion of fossil fuels. The world is gradually beginning to recognize the issues and diseases caused by traditional fuels [34, 35].

The sustainable production of biofuel is a valuable tool in stemming climate change, boosting local economies, particularly in less developed parts of the world, and enhancing energy security globally. Therefore, the exploration of novel, renewable, environmentally friendly, clean, reliable, and economically feasible energy resources is an urgent need today. Biofuels offer a competitive alternative to conventional fossil fuels in the transportation, industrial, residential, commercial, and power generation sectors. Compared to nonrenewable energy sources, biofuels enhance energy security, reduce greenhouse gas emissions, and support rural livelihoods by creating jobs and market opportunities for home harvests. They also offer some commercial benefits, such as a more diverse fuel mix, greater sustainability, and the potential to generate a large number of jobs in rural areas. In addition, they have the potential to boost the economy (revenue), increase industrial investments, and develop the farming and agricultural sectors. Biofuels are good alternatives for climate change because they reduce greenhouse gas emissions and air pollution, are easy to biodegrade, improve combustion efficiency and carbon sequestration, increase supply reliability, use fewer fossil fuels, and are readily available and distributed [31, 34, 36–38].

8.2.2 Role of carbon-based catalysts in biofuel synthesis

In the process of producing biofuels, catalysts are crucial. Researchers are working to create catalysts from renewable resources, such as biomass, in an effort to make the process completely sustainable. Transesterification is the simplest and most economical method of producing biodiesel. Although either an acid or a base can catalyze the reaction, transesterification generally works best with base catalysts (figure 8.2) [39]. The conventional methods of producing biofuels use homogeneous base catalysis for the transesterification reaction. Although homogeneous base catalysis has a high biodiesel yield, favorable operating conditions, and a quick reaction rate, there are a number of barriers that prevent the process from being profitable [40, 41]. Heterogeneous catalysts, such as carbon-based catalysts, have now been introduced and employed in biofuel synthesis.

Carbon-based catalysts have been presented for use in various processes. Because of their favorable properties, which include high surface area, low material cost, and thermal stability, carbon-based materials are regarded as the best catalysts. The carbon surface can be readily functionalized with acids or bases to create them; in other situations, carbon material has reportedly been utilized as a support. Furthermore, much of the waste generated by various industrial activities can be converted into carbon. As a result, using it as a catalyst improves the process of

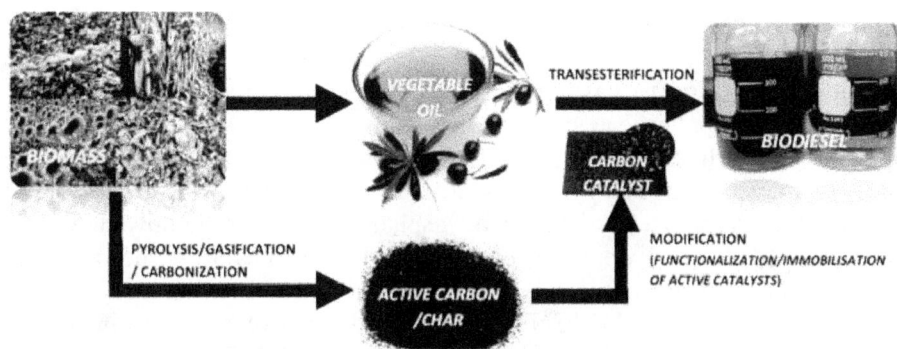

Figure 8.2. Schematic representations of the preparation of carbon-based catalysts and their utilization for biodiesel production. Reprinted from [39], Copyright (2014), with permission from Elsevier.

producing biodiesel [39]. Carbon nanoparticles, including sulfonated CNTs, supported carbon materials, and acid- and base-activated carbon, have been shown in numerous studies to have great potential as biofuel production catalysts. Ballotin et al [42] explored a catalyst made of carbon nanostructures encased in amorphous carbon. The esterification reaction between oleic acid and methanol was catalyzed in this work by a solid acid catalyst based on several nanostructures, such as graphene, nanotubes, nano-onions, and nanographite, embedded in an amorphous carbon matrix. The resulting materials demonstrated 95% conversion rates when utilized as catalysts for the esterification of oleic acid with methanol.

The overall economy of biofuel production depends mainly on two crucial factors: (i) the feedstock and (ii) the catalyst. The production of biodiesel (figure 8.2) is significantly more environmentally friendly when a carbon-based catalyst is used.

Carbon-based catalysts have several advantages, including uniform particle dispersion, large surface area, high thermal stability, ease of synthesis, cost-effectiveness, ease of deployment, and reusability. Generally, particulate heterogeneous catalysts can also be reused because they are easily removed from products following a reaction, resulting in less waste and energy consumption. Other advantages of these catalysts over their predecessors include higher site density, enhanced catalytic activity, consistent operation, and affordable pricing in an environmentally favorable setting [43–45]. To increase their catalytic capabilities through synergistic effects, they are also commonly used as supporting frameworks or skeletons for nanoparticulate catalysts. Their unique structures allow for effective particle separation, distribution, and anchoring while preventing aggregation and sintering. For instance, they can give metal-based electrocatalysts a stable structure, allowing biofuels to be converted into sustainable energy sources. Thus, these catalysts make the synthesis of biofuel sustainable [46, 47].

8.2.3 Case studies in biofuel production

A number of studies have explored the use of carbon-based catalysts in biofuel production. The pyrolysis of a glucose-impregnated polymer matrix, followed by sulfonation, produced a new sulfonated carbon composite solid acid. Compared to the previously documented carbon-based catalyst made by sulfonating pyrolyzed sugar, it was reported to have superior reusability and free fatty acids (FFAs), namely acetic and palmitic acid. The catalyst also demonstrated greater esterification activity [48]. In another study [50], the transesterification of methanol with cottonseed oil was catalyzed by a carbon-based solid acid (CBSA) catalyst created by sulfonating carbonized vegetable oil asphalt. Sulfonated multiwalled carbon nanotubes (s-MWCNTs) were also prepared and used to catalyze the same transesterification as the asphalt catalyst. The asphalt-based catalyst showed higher activity than that of the s-MWCNTs for the production of biodiesel. In the study reported in [49], partially carbonized de-oiled canola meal (DOCM), a by-product of processing canola seeds, was sulfonated to produce a carbon-based acid catalyst. The partially carbonized material and DOCM were treated with strong sulfuric acid to create four different kinds of carbon catalysts. The catalyst has enough acidic sites in the form of sulfonate groups to support the esterification process, as demonstrated by spectral analysis and elemental studies.

Three solid acid catalysts were prepared by the sulfonation of carbonized vegetable oil asphalt and petroleum asphalt in another study [50]. When a model waste oil feedstock was utilized, a novel method that combined reaction and separation significantly enhanced the conversion of cottonseed oil (triglyceride) and FFAs. The catalyst with the highest catalytic activity was the one made of vegetable oil asphalt. By covalently attaching aryl radicals containing sulfonic acid to the surface of mesoporous carbons, sulfonated ordered mesoporous carbons (OMCs) were created [51]. The resultant materials exhibited consistent and very effective catalytic performance in the manufacture of biodiesel when compared to existing solid acid catalysts, with the highest conversion rate reaching 73.59%. High-grade biofuels were produced in a study [52] by pyrolyzing waste clay oil using a catalyst based on seaweed char. The seaweed carbon-based catalyst had a good decarboxylation effect, according to the results. Liao *et al* [53] used sulfur-modified carbon nanotubes (S-CNTs) as supports, and a series of Pd–Co bimetallic catalysts were readily produced using the impregnation approach. Because of the support-enhanced adsorption effect and the efficient synergy between the highly disseminated metallic Pd and the Co9S8 nanoparticles, the bifunctional Pd-Co9S8/S-CNT demonstrated impressive hydrogenolysis/hydrogenation performance. It was found that 96.3% Kraft lignin was chemically activated using phosphoric acid, sulfuric acid, and pyrolysis to create a solid acid catalyst. It was employed to catalyze the one-step conversion of untreated Jatropha oil to biodiesel and the esterification of oleic acid. The result showed that an esterification rate of 96.1% produced the maximum catalytic activity, and under ideal circumstances, the catalyst could be employed three times with minimal deactivation.

By directly sulfonating the remaining lignin from *Xanthoceras sorbifolia* Bunge hulls, a lignin-derived carbonaceous catalyst (LCC) was created as a solid acidic catalyst. Using the LCC, biodiesel was produced from acidified soybean soap stocks. When sulfuric acid and LCC were compared for catalytic activity, LCC outperformed sulfuric acid with the same active group (single-bond SO_3H) by a factor of 3.5. Furthermore, LCC had a high FFA conversion rate and could be reused at least three times [55]. The esterification of oleic acid and methanol, a crucial step in the manufacture of biodiesel, was carried out using sulfonated carbons as solid acid catalysts. With a turnover frequency of eight times that of Amberlyst-15 and a rate constant three times that of Amberlyst-15, the improved carbon catalyst demonstrated significantly higher activity than the conventional solid acid [56].

Carbon-based nanocatalysts are also a major type of carbon-based catalyst. The primary component of carbon-based nanomaterials is carbon atoms arranged in special structures with nanoscale dimensions that provide remarkable qualities used in a range of applications. The benefits of carbon-based nanocatalysts include high thermal stability, a metal-free nature, and simplicity of synthesis. Carbon-based nanocatalysts are preferred due to their novel characteristics and significant output [46]. Macina *et al* [57] demonstrated that glycine and citric acid mix hydrothermally to create a heterogeneous carbon dot catalyst for canola oil transesterification, which yields biodiesel. With the addition of 1 wt% of catalyst, they reported biodiesel conversion rates of 97% or higher at 150 °C. Furthermore, they showed that the nanocatalyst was stable and reusable, maintaining catalytic efficiency for a minimum of five reaction cycles. The impact of employing amide-functionalized CNTs on the ignition delay of blends of these nanomaterials (50 and 100 ppm) and Colombian commercial diesel (90% petroleum diesel and 10% palm-oil biodiesel) was also examined for the first time by Rentería *et al* [58].

Sulfonated acid-functionalized carbon catalysts for biofuel synthesis have also been documented in the research literature. They can catalyze the esterification of FFAs (found in oil) or, similarly to concentrated H_2SO_4, concurrently catalyze esterification and transesterification reactions to produce biodiesel. In [59], a solid Brønsted acid of amorphous carbon with SO_3H, COOH, and phenolic OH groups was investigated as a catalyst for the generation of biodiesel [59]. The carbon material exhibited remarkable catalytic performance for the transesterification of triolein, maintaining high catalytic activity even in the presence of water. These results suggest that this catalyst can directly convert crude vegetable oils composed of triglycerides, free higher fatty acids, and water into biodiesel with minimal energy consumption. This study also looked at the possibility of producing ecologically friendly biodiesel using the esterification of oleic acid and the transesterification of triolein over amorphous carbon carrying SO_3H, COOH, and phenolic OH groups. A carbon material known as a sugar catalyst is a novel kind of amorphous CBSA catalyst made of nanographene containing SO_3H groups [44, 59]. The process of creating such a catalyst using common, low-cost sugars is described in [60]. This high-performance catalyst, which is recyclable and made of stable sulfonated amorphous carbon, is significantly more active than other solid acid catalysts that have been tested for the manufacture of biofuel.

The esterification and transesterification reactions required to produce biodiesel have been shown to benefit greatly from the exceptional catalytic activity of a novel class of carbonaceous catalysts that are produced by partially carbonizing carbohydrate sources such as D-glucose, sucrose, starch, and cellulose and then sulfonating them. CBSA is also a possible heterogeneous catalyst. A number of CBSAs with diverse origins have been investigated, including D-glucose, sucrose, starch, cellulose, and vegetable oil asphalt that were charred and sulfonated. CBSAs are formed from either carbohydrate or biomass precursors [43]. Compared to commercially available solid acid catalysts, they have demonstrated superior activity for both the transesterification of triglycerides and the esterification of FFAs. In Villa *et al* [61], multiwalled CNTs grafted with different amino groups exhibited remarkable stability and activity as fundamental catalysts for triglyceride transesterification. Three distinct amines (triethylamine, ethylamine, and pyrrolidine) were grafted to create three catalysts to link the amines' varying basicity with catalytic activity. It was concluded that grafting amino groups onto CNTs is an appropriate technique for producing stable and active basic catalysts that can be used in liquid-phase processes [61].

Nitrogen-functionalized CNTs were synthesized by grafting amino groups to the surface of the nanotubes. In the base-catalyzed liquid-phase transesterification of glyceryl tributyrate with methanol, a typical reaction for the synthesis of biodiesel, the nanotubes demonstrated encouraging outcomes [62]. In [63], a series of deprotonation/carbometalation and electrophilic substitutions were used to create an NEt_3-graphene solid basic catalyst. Xu *et al* [64] reported a catalyst made of Pd nanoparticles supported on mesoporous N-doped carbon, which was shown to be highly active in promoting biomass refining. A high-nitrogen-content mesoporous carbon material that showed high activity in stabilizing Pd nanoparticles was produced using silica nanoparticles as a hard template and a task-specific ionic liquid (3-methyl-1-butylpyridine dicyanamide) as a precursor. The resulting Pd@CN0.132 catalyst showed very high catalytic activity in the hydrodeoxygenation of vanillin at low H_2 pressure under mild conditions in aqueous media. Under ideal reaction conditions, high biodiesel yields of 86% and 91% were observed for acidic and basic catalysts, respectively. For up to eight catalytic cycles, both types of catalysts demonstrated exceptional recoverability [65].

8.2.4 Biodiesel from biomass

Due to its potential, biomass is a viable solution for meeting demand and ensuring a sustainable supply of energy and fuel in the future. One potential path toward the effective use of biomass resources is the modernization of biomass technologies, which will result in more efficient biomass production and conversion [32, 43]. The process of turning biomass into biofuels has advanced significantly in recent years. Catalysis is a crucial element of the biofuel processing chain, essential for enhancing both the yield and quality of biofuels, and it holds much promise because it is unquestionably one of the best methods for obtaining the required chemical feedstocks from biomass. In a study [66], de-oiled *Jatropha curcas* (JC) seed cake waste was used to create a CBSA catalyst. The high FFA concentration of the JC oil was subsequently reduced to an appropriate level (<4 mg KOH g^{-1}) for the synthesis of biodiesel by esterifying it using the catalyst. In terms of the reaction time required to obtain the maximum conversion yield, the catalyst was also shown to perform better than the sulfuric acid catalyst that is typically utilized [66].

Bastos *et al* utilized an agro-industrial waste, murumuru kernel shell, as the precursor biomass to create acid biochar, which was then used as a catalyst to produce biodiesel made from jupati oil. The catalyst was created by carbonizing the shells of murumuru kernels and then sulfonating the resulting material in concentrated sulfuric acid. The feasibility of using this industrial waste product as a precursor for the sustainable production of an effective sulfonated carbon catalyst for the synthesis of biodiesel was demonstrated by the study's findings on ester content and reuse [67]. The possibility of using biomass carbon obtained from rice husks, *Moringa oleifera* seeds, and lipid-extracted marine algae (BM) with sulfuric acid groups for the generation of environmentally friendly biodiesel was examined by Rana *et al* [68]. Compared to other solid acids and a sulfuric acid catalyst, SO_3H-RH exhibited comparatively higher catalytic activity and stability in the synthesis of biodiesel [68].

In another study [69], de-oiled microalgal biomass was carbonized and then sulfonated to create a new CBSA catalyst. The study established how the catalyst's surface acidity and the conversion of FFAs were affected by the catalyst synthesis parameters, including the carbonization temperature, sulfonation time, and H_2SO_4 concentration. Up until the fourth cycle, the catalyst showed good catalytic activity. A further study [70] compared the catalytic performance of carbon-based acid catalysts derived from papaya seed, empty fruit waste, and corncob biomass waste, which were used to produce biodiesel through the esterification reaction of palm fatty acid distillate and methanol to assess the use of biomass wastes as catalyst supports. The sulfonated activated carbon catalyst made from corncob waste had the highest yield and FFA conversion rates of 72.09% and 93.49% for palm fatty acid distillate and methanol, respectively [70].

The production of a novel ordered mesoporous carbon-based catalyst (MCC) with superacid sites from waste microalgal biomass was investigated. Biochar was produced by partially carbonizing the waste microalgal biomass. The resulting biochar was then subjected to a sulfonation process with concentrated sulfuric acid. The conversion of linseed oil to biokerosene, which can be utilized as a biokerosene component for blending with aviation fuels, effectively demonstrated the catalytic activity [71]. Dhawane *et al* [72] used a carbonaceous heterogeneous catalyst generated from *Delonix regia* (flamboyant) pods to demonstrate the parametric effects on the generation of biodiesel from *Hevea brasiliensis* oil (HBO). The contribution factor showed that the methanol-to-oil ratio and catalyst loading had the most significant effects on the yield of biodiesel out of the four parameters taken into consideration. It was also predicted that the cost of preparing a carbonaceous catalytic support could be substantial. Therefore, pea pod-derived carbon can be regarded as an efficient catalyst for the manufacture of biodiesel, and HBO may be regarded as a good feedstock [72].

In order to produce biodiesel from the inedible feedstock JC oil (JCO), Ruatpuia *et al* thoroughly tested the batch reproducibility of a sulfonic acid-functionalized carbonaceous material as a catalyst. The results confirmed that the biodiesel product had a chemically acceptable composition, that the catalyst could be reused, and that the use of microwave heating significantly improved performance. The fifth reuse cycle showed a JCO conversion rate of 83.0 ± 0.8% [73]. Using a carbonization–sulfonation process, a series of CBSA catalysts were made from bagasse and used to catalyze the esterification of oil with methanol to create biodiesel. The catalysts made from bagasse using diluted acid hydrothermal treatment performed the best, according to Zhang *et al* [74]. In another study, murumuru kernel shell, an agro-industrial waste, was used as the precursor biomass in the synthesis of an acid biochar that was employed as a catalyst in the production of biodiesel originating from jupati oil [75].

8.3 Catalytic processes in CO_2 conversion

8.3.1 Mechanisms of CO_2 conversion

The conversion of CO_2 plays two roles by contributing to the reduction of greenhouse gases while mitigating the energy crisis. Since CO_2 is known to be the most thermodynamically stable of the carbon species, some studies state that its conversion into renewable energy requires high energy consumption [78]. While this is true for its reaction with O_2, CO_2 can serve as a carbon and oxygen source, thus reacting with hydrogen-containing compounds, which can supply enough energy for the conversion to take place without requiring high energy consumption [76].

CO_2 is utilized as an important building block in the synthesis of chemicals such as methanol, synthesis gas, carbonates, and carboxylic acids [77]. One example is the dry reforming of methane, which utilizes two greenhouse gases—CH_4 and CO_2—in the production of syngas (H_2 and CO) [78]. The syngas can be further used to produce methanol, ammonia for fertilizers, diesel, and gasoline-grade hydrocarbons, as well as to generate heat and power (Havilah *et al* 2022). Products that are potential alternatives to petroleum-derived fuels/chemicals include CH_4, C_2H_6, CH_3OH, C_2H_5OH, etc.

Approaches to CO_2 capture and control have been grouped into two categories, namely CO_2 capture and storage (CCS) processes and CO_2 capture and utilization (CCU) processes [79]. The CCU process results in the utilization of the CO_2 after capture and storage [80]. The combination of both groups is called carbon capture utilization and storage (CCUS), which was recognized by the Paris Climate Conference and is expected to remove about 19% of global CO_2 by 2050 [81]. Barker–Rothschild *et al* 2024) highlighted a third group, namely reactive carbon capture (RCC), which involves the immediate conversion of CO_2 into other products, thereby improving energy efficiency by eliminating the thermal energy required for CO_2 release.

CCU technologies include chemical and physical absorption, adsorption, membrane separation, electrochemical, thermochemical, and photocatalytic approaches [81, 82]. Absorption, which involves the dissolution of CO_2 in a solvent (which can cause a reaction with the solvent), has been one of the most widely used techniques; however, the cost of the energy required for solvent regeneration is relatively high due to solvent degradation [83]. Adsorption technologies offer advantages in their ease of regeneration, low energy consumption, and the absence of hazardous by-products [84]. Electrochemical processes, although complex, are advantageous for

CO_2 reduction due to their scalability, controllable processes, and minimal chemical consumption [85].

8.3.2 Case studies illustrating CO_2 conversion using carbon nanomaterials/composites

Carbon has been identified as one of the Earth's most abundant and widely available elements [86]. It has distinct inherent characteristics when compared to noble metals, such as a high surface area with defect sites, acid and base resistance, high thermal stability, porous structure, and environmental friendliness [79]. These characteristics, alongside its chemical inertness, make carbon an effective material for CO_2 reduction, alone or coupled as a catalyst support with other semiconductors [86]. Dissanayake *et al* [84] used biochars produced from mesquite wood chips and chicken manure as adsorbents. Tests showed that 70% of their maximum adsorption capability could be reached in 10 min, with the most optimum formulation exhibiting 78.6% of its maximum adsorption capacity. Shao *et al* [87] synthesized nitrogen and sulfur co-doped porous carbon and obtained a CO_2 uptake of 3.54 m mol g^{-1}.

In a study by Zhang *et al* [88], the electrochemical reduction approach was utilized for the direct conversion of CO_2 to CH_4 using graphene quantum dots (GQDs). This was achieved through the functionalization of electron-donating groups, such as OH and NH_2, on the GQDs to promote CO_2 conversion. The results also showed that the functionalization of GQDs with electron-withdrawing groups such as COOH and $-SO_3$ inhibited the CO_2RR. Yue *et al* [89] ranked the electrochemical activity of some carbon materials in the electrochemical reduction of CO_2. Their study included six carbon-based materials, namely single-layer graphene (SLG), graphite paper (GP), buckminsterfullerene (C_{60}), glassy carbon (GC), boron-doped diamond (BDD), and nitrogen-doped reduced graphene oxide (NRGO). In a comparison of their Faradaic efficiency for CO (FECO) potentials, NRGO had the highest potential (-0.9 V), followed by BDD (-1.4 V), GC, C_{60} (-1.6 V), and GP (-1.7 V).

Ye *et al* [90] designed a novel nickel-nitrogen-modified porous carbon/carbon nanotube hybrid (Ni–NPC/CNTs) electrocatalyst for the selective reduction of CO_2 to CO. The results showed that the integration of amorphous carbon and CNTs improved the surface area and electron transfer efficiency of the electrocatalyst, while its geometry also favored mass transfer on the surface. This resulted in an excellent electrochemical reduction of carbon dioxide, with 94% FECO.

8.3.3 Photocatalytic CO_2 reduction using carbon-based nanomaterials/composites

Photocatalysis has received increased attention because of the crucial role it plays in environmental conservation and clean energy. It is based on the concept of natural photosynthesis and employs a semiconductor material acting as a photocatalyst with a reducing agent to convert CO_2 to useful chemicals under light irradiation [91].

The photocatalytic process involves three key steps [92]:

    I. Absorption of light energy by the photocatalyst and the generation of photoexcited electron–hole pairs

    II. Separation and transfer of the charge carriers

    III. Chemical reactions between the surface species and the charge carriers

Table 8.1. Studies that reported the carbon-based photocatalytic reduction of CO_2.

Catalyst	Method	Product	Evolution	Time	Irradiation source	References
ZnS/CdS/ reduced graphene oxide (rGO)	Photocatalysis	CO	38.77 μmol g^{-1}	4 h	UV–Vis	[93]
TiO$_2$/g-C$_3$N$_4$ nanosheet	Photocatalysis	CO	1.96 μmol g^{-1} h^{-1}		UV–Vis	[94]
WO$_3$/g-C$_3$N$_4$	Photocatalysis	CO	58.4 μmol g$_{cat}^{-1}$	4 h	UV	[95]
		CH$_4$	41.47 μmol g$_{cat}^{-1}$			
Ag-TiO$_2$/rGO	Photoelectrocatalysis	CH$_3$OH	85 μmol l^{-1} cm^{-2}	210 min	UV–Vis	[96]
CoAl-LDH/ CeO$_2$/rGO	Photocatalysis	CO	5.5 μmol g^{-1} h^{-1}		UV	[97]
ZnO/Au/ g-C$_3$N$_4$	Photocatalysis	CO	689.7 μmol m^{-2}	8 h	UV–Vis	[98]
g-C$_3$N$_4$/Pd/ MoO$_{3-x}$	Photocatalysis	CO	3.92 μmol g^{-1}	4 h	NIR	[99]

Table 8.1 shows some studies where carbon-based materials were used for the photocatalytic conversion of CO_2.

8.3.4 Use of carbon in energy-efficient catalysis

While a range of inorganic semiconductors such as TiO_2, ZrO_2, and CdS have been investigated and utilized for the photocatalytic reduction of CO_2, they are not suitable for visible light activation due to their unaligned valence and conduction bands, leading to large bandgaps that are unable to absorb visible light [100]. While visible light illumination is the key driver for photocatalytic reactions, the photocatalytic efficiency is mainly determined by the band structure of photocatalysts that can absorb visible light [101]. TiO_2 is a catalyst that is utilized in photoreduction, but its characteristic disadvantage of a large bandgap, resulting in the recombination of photogenerated electron–hole pairs and low quantum yields, makes it unsuitable for visible light activation. However, 45% of the solar spectrum consists of visible light, and it is important to harness this available energy resource [102].

Researchers have developed nanocomposite combinations with carbon structures (graphene, fullerene, etc.) and semiconductors such as TiO_2 for visible light activation in the conversion of CO_2. In their study [103], Rodríguez *et al* created a reduced graphene oxide (rGO)/TiO_2 and rGO/TiO_2/Cu nanocomposite for the CO_2 photoreduction reaction. The rGO acted as a sensitizer, improving the visible light activity of the composite while also acting as an electron reservoir to prevent electron–hole recombination. The Cu also acted as an electron trap, providing an active site for CO_2 activation and dissociation. Nanoporous reduced graphene deposited on the surface of CdS on a porous aluminum support photocatalyst resulted in the generation of 153.8 mol g^{-1} hr^{-1} of CH_3OH when the photocatalytic reduction of CO_2 was carried out using a compound parabola as a solar reflector [104]. An example of a non-metallic photocatalyst is graphitic carbon nitride ($g-C_3N_4$), known for its optical properties (bandgap $=$ 2.7 eV), environmental friendliness, and low cost. In this catalyst, the oxidizing sites tend to be nitrogen atoms, while the carbon atoms are the reducing active sites [105]. However, this catalyst is still characterized by high recombination rates in photoinduced electron carriers, resulting in poor performance. For this reason, studies have shown that coupling it with other materials with a lower bandgap results in better visible light absorption effectiveness [106].

In another study [107], a Z-scheme composite catalyst, $Ag_3PO_4/g-C_3N_4$, was synthesized for the conversion of CO_2 to fuel under visible light irradiation. Ag_3PO_4, a semiconductor, has been reported to exhibit visible-light-driven photocatalytic activity. The $Ag_3PO_4/g-C_3N_4$ photocatalyst showed a conversion rate 6.1 times better than that of a $g-C_3N_4$ photocatalyst in visible light. The presence of a plasmonic Ag compound induced strong absorption of the incident light and promoted the separation of the photoinduced electron-hole pairs. This phenomenon is referred to as surface plasmon resonance (SPR) [108].

Lu and co-workers [109] coupled Z-scheme 2D $BiVO_4$ and ultrathin $g-C_3N_4$ nanosheets. These were synthesized through a thermal polymerization technique and a hydrothermal method. While the Z-scheme arrangement improved charge transfer, the efficiency was further enhanced by the geometry of the 2D/2D nanosheets due to an increase in the interfacial contact area and an improved surface area, resulting in more active sites for the photocatalytic reaction to take place. The results showed optimal conversion of CO_2 to CH_4 and CO, which were 4.8 and 4.4 times those achieved using $g-C_3N_4$.

8.4 Green energy production using carbon-based materials

In recent years, different forms of carbon and their derivatives have seen a high level of application in many technologies, including energy storage [110], nuclear reactors [111], sensing [112], thermal cooling in electronics [113], and solar energy conversion [114]. These broad applications of carbon-based materials are due to their many beneficial properties, such as good conductivity, mechanical flexibility, good stability, low production cost, and environmental friendliness, allowing for their application in green energy production [115]. In solar energy conversion, carbon-based materials can serve numerous positive roles in different aspects of device functionality, such as boosting performance, improving stability, enhancing sustainability, and widening the application scope. Solar energy can be harnessed using photovoltaic (PV) technology, which directly converts sunlight into electricity, and solar thermal systems, which produce heat as the end product. In addition, integrated systems have also evolved, which simultaneously use both PV and solar thermal systems for efficient solar energy utilization.

8.4.1 Photovoltaics

The versatility of carbon-based materials' properties, preparation routes, and ease of morphological tuning has allowed for their application in various PV systems, such as solar cells based on silicon (Si), perovskite, organics, and dye sensitization, playing vital roles in boosting performance. In addition, they are also applied in the thermal cooling of PV modules for optimal performance. In dye-sensitized solar cells (DSSCs), carbon-based materials play multiple roles at different layers of the device architecture. A clear demonstration of this was reported by Islam et al [116]. They developed a DSSC that utilized p-graphene as the front electrode instead of the conventional fluorine-doped tin oxide (FTO) and indium tin oxide (ITO)for reduced cost, while graphene oxide served as the charge transport layer rather than the widely used TiO_2. In addition, [6,6]-phenyl-C_{61}-butyric acid methyl ester (PCBM), a derivative of C_{60}, was used as a composite material with poly[N-9'-heptadecanyl-2,7-carbazole-alt-5,5-(4',7'-di-2-thienyl-2',1',3'-benzothiadiazole)], poly[[9-(1-octylnonyl)-9H-carbazole-2,7-diyl]-2,5-thiophenediyl-2,1,3-benzothiadiazole-4,7-diyl-2,5-thiophenediyl] (PCDTBT) and served as a solid electrolyte for the DSSC. The solar cell attained an efficiency of 13.38%, which was higher than that of many reported conventional DSSCs and above the 10% threshold needed for commercialization [116]. Recently, carbon fibers derived from cotton, Mo-doped carbon rods, and MoS_2 were used to develop transparent, stable, cost-effective, and high-performance counter electrodes that competed with platinum in a DSSC. The electrodes produced low charge transfer resistances of 9.45 and 6.43 Ω, respectively, and achieved slightly higher efficiency compared to platinum (Pt) when applied in a DSSC [117].

Silicon-based cells have dominated the PV market for many decades. However, the cost of PV power is still much higher than that of power produced using conventional fossil fuel sources. The integration of carbon-based materials is vital in bringing down the cost of Si-based solar cells while boosting device stability and/or efficiency. For example, forming heterostructures that combine carbon-based materials with Si has shown great potential for enhancing the device performance of Si-based solar cells. A heterojunction made by combining graphene and Si in a solar cell, which also incorporated poly(methyl methacrylate) (PMMA) as an antireflective surface, yielded a high-performance device, attaining a maximum power conversion efficiency (PCE) value of 13.7% [118]. In another study, the introduction of graphene oxide as an interlayer in a poly(3,4-ethylenedioxythio-phene) polystyrene sulfonate (PEDOT:PSS)/n-Si solar cell enhanced the lifetime of minority charge carriers by seven times and increased the device PCE value by 3.1%. The interlayer served multiple functions of passivating oxygen vacancies at the interfaces while promoting more effective separation and collection of charge carriers [119]. Other studies have employed single-walled CNTs (SWCNTs) combined with an organic passivation scheme, such as Nafion, which served as a conductive and passivating layer, attaining device efficiencies of up to 23% [114, 120]. This scheme seeks to replace the use of metal contacts in Si-based solar cells (which often serve as recombination centers for photogenerated charge carriers) with carbon-based materials, thereby reducing costs.

Organic solar cells (OSCs) have evolved over the years as a promising PV technology for the green production of energy. This is because OSCs offer new PV technologies such as flexible modules with diverse applications in buildings, power mobility, etc. However, the efficiency of OSCs is still significantly below that of their Si-based counterparts. Carbon-based materials can be applied in nearly all the layers of OSC architecture for performance enhancement. In a theoretical study, CNT-based materials were used to replace all the materials in the different layers of a bulk-heterojunction (BHJ)-based OSC in a robust investigation that used DFT calculations and other modeling/simulation tools. A notable conversion efficiency of 19.63% was reported for a CNT-based OSC. This study showcases CNTs as a promising material for the advancement of OSCs and provides relevant information for future experiments [121]. In an experimental study, carbon quantum dots (CQDs) were incorporated into PEDOT:PSS, a widely investigated material used as a hole transport layer in OSCs, which enhanced the cells' contact and charge transport properties, improving their PCE by 3.9% [122]. In another remarkable application of carbon-based materials in green energy generation, an isotropic fullerene was attached to a Y-series electron acceptor in an OSC. The resulting OSC enhanced exciton dissociation and limited energy losses due to non-radiative processes, producing a PCE of 15.02%. When included in a ternary blend setting, a conversion efficiency of 19.22% was achieved, which is among the best values reported for OSCs [123].

Meanwhile, perovskite solar cells (PSCs) are an evolving technology that has attracted extensive research effort since the first report of an efficient solid-state PSC in mid-2012 [124]. Perovskites possess beneficial electrical and light-harvesting properties that make them stand out. A PCE value of over 25% has been reported for PSCs [125], which competes with the values reported for the commercially dominant Si-based solar cells. Moreover, PSCs offer the competitive advantage of being made from cheap and easily processible materials, leading to low-cost PV devices compared to their Si-based counterparts. The key drawbacks of PSCs are their instability and toxicity. The first involvement of carbon-based materials in PSCs was reported in 2013, when carbon black/graphite was employed as a counter electrode in a $CH_3NH_3PbI_3/TiO_2$ PV device, and a PCE value of 6.64% was attained [126]. Since then, carbon-based materials have been used in the further development of PSCs [127, 128].

One of the key challenges of PV cells is the generation of heat due to localized heating and hot spots, which can significantly limit their performance [129]. The cooling of PV modules is relevant for attaining optimal device performance. Carbon-based materials are also playing significant roles in the cooling of PV systems. Graphene-nanoparticle-based nanofluids were employed in the cooling of PV solar panels, and the performance was compared with that of water as a coolant. The graphene-based system showed superior cooling capabilities and achieved a cooling efficiency 42% greater than that of water at the highest flow rate [130]. In a recent study, Karagan *et al* examined the cooling efficiency of nanofluids based on hybrids of graphene oxide and CNTs with CuO, respectively. CNT-CuO composite-based nanofluids yielded the best output, exhibiting a 12.07% improvement in power output and a 31.2% energy efficiency [131].

8.4.2 Solar thermal systems

In another solar energy conversion technology, the radiation from the sun is converted into heat by a solar thermal system, which can then be used for many purposes, such as heating and cooling in homes, drying, and the desalination of water. This approach can utilize different kinds of solar collectors [132]. However, numerous challenges still exist around boosting solar thermal conversion efficiency and optimizing heat transfer and storage. In attempts to deal with these challenges, many carbon-based materials have been usefully employed in different types of solar thermal systems.

Hierarchical vertically grown graphene foam was developed using plasma-enhanced chemical vapor deposition (PECVD) and applied as the heating medium in a solar vapor conversion system. An external solar thermal conversion efficiency of about 93.4% was achieved, and 90% efficiency was obtained in solar vapor conversion for seawater desalination [133]. In addition, nanofluids based on graphene oxide and its binary composites with ZnO and FeO were compared in a direct absorption solar thermal application [134]. The pure graphene oxide out-performed all the composite systems, exhibiting 92% solar conversion efficiency due to its superior physicochemical properties, conductivity, and stability when dispersed in fluids. Several performance enhancements have also been reported for other carbon-based materials, such as CNTs [135] and carbon fibers [136], in various solar thermal applications.

8.4.3 Integrated systems

Integrated solar conversion systems that combine PV and solar thermal technologies have also gained wide adoption in recent years. The photovoltaic units of these integrated systems convert solar energy into renewable electricity, while the thermal system produces heat. The thermal unit can simultaneously provide cooling to the PV module, boosting its PCE and durability while producing heat for useful applications. Carbon-based nanomaterials have also been instrumental in the development of integrated PV/solar thermal systems. A report on graphene-based ionanofluids employed as a heat transfer medium in a hybrid PV/thermal system described a 5% increase in energy efficiency relative to the sole use of the ionic liquid [137]. In another demonstration of the effective use of carbon-based materials in PV/thermal systems, a hybrid nanofluid of multiwall carbon–silicon carbide was simulated. The results showed that its average thermal and electrical energy efficiencies were 56.55% and 13.85%, respectively [138].

8.5 Carbon nanomaterials in hydrogen production (water splitting and biomass gasification)

Hydrogen (H_2), a clean fuel that yields water as a by-product after combustion, can be produced in numerous ways, such as fossil hydrocarbon reforming, steam reforming, PEC water splitting, coal gasification, and biomass gasification [139]. PEC water splitting and biomass gasification are among the top technologies that hold great promise for the sustainable green production of H_2. PEC water splitting uses energy from the sun to electrolyze water, producing hydrogen. Meanwhile, biomass gasification is considered one of the best routes for obtaining a high yield of H_2 from biodegradable feedstocks.

8.5.1 Photoelectrochemical water splitting

The basic concept of PEC water splitting involves the use of two key electrodes. A photocatalyst, which can be a photoanode or photocathode, and a counter electrode are combined with an electrolyte to drive the electrolysis of water, using solar radiation as the energy source [140]. In the last few decades, many materials such as metal oxides, nitrides, and metal sulfides and oxides have been investigated for use as photoelectrodes in PEC water splitting [141]. However, most of these materials either suffer from poor charge transport properties or are unstable in solution. The superior conductive properties of carbon-based materials and their excellent stability in aqueous solution make them suitable for advancing further photoelectrode development. They are cheap alternatives to the conventional Pt that is often used as a counter electrode in PEC water splitting [142].

In the advancement of materials development for PEC water splitting, several carbon-based materials have been used directly [143] or to modify other photoelectrodes [144, 145]. The incorporation of CNTs on ZnO to form a heterostructure has been reported to produce a 33-fold increase in photocurrent density in a PEC cell for water splitting [144]. Moreover, a highly efficient photoelectrode for PEC water splitting that consisted of a carbon-rich C_3N_4 and TiO_2 composite yielded a photocurrent of 4.30 mA cm^{-2} at 0.6 V vs Ag/AgCl (compared to 0.38 mA cm^{-2} for bare TiO_2 at the same potential), a hydrogen evolution rate of 26.51 μmol h^{-1}, and a faradic efficiency of 88.12% [145]. Carbon-based materials such as C_3N_4 have been used as a sole photocatalyst and showed impressive stability over a pH range of 0–13 [143]. Carbon-based materials can also be used to modify the surface of other photoelectrodes to boost their charge separation properties and improve their stability in an electrolyte [146]. Furthermore, carbon-based materials such as graphite rods [142] have been employed as counter electrodes in a PEC cell and can be further explored as a potential replacement for the conventional Pt often used for the same purpose in PEC water splitting.

8.5.2 Biomass gasification

Generally, over 40% of the total energy content of biomass is contained in H_2. Therefore, biomass is rich in H_2 that can be harnessed using thermochemical, biological, and electrochemical methods [147]. Biomass gasification is considered an efficient approach for the green thermochemical conversion of biodegradable feedstocks to produce H_2 because of its potential for high yield [147, 148]. The use of a catalyst in the biomass gasification process is vital to improving the selectivity of H_2 and its production and thus advancing this technology. Many conventional catalysts used in the gasification of biomass, such as ores, alkaline earth metals, and noble metals, are either inefficient or expensive [148]. Carbon-based materials are now employed in the development of efficient methods for advancing hydrogen production via biomass gasification.

Carbon-based materials can be used as composites with conventional catalysts, such as Pt and Ru, to improve performance and reduce costs. In a recent study [149], a hierarchical porous carbon-based catalyst modified with $NaHCO_3$/Pt was used in the hydrothermal gasification of lignocellulosic biomass. The gasification process yielded an impressive maximum H_2 production rate of 23.81 ml H_2 mg^{-1} Pt. The impressive performance was largely linked to the porous characteristics of the hydrocarbon used, coupled with the presence of sodium species in the catalyst [149]. Elsewhere, another carbon-based material, graphene, served as the support for the loading of Pt and was gainfully applied as a catalyst in the gasification of biomass compounds [150]. CNTs have also been confirmed to be suitable supports for the loading of Ru in catalyst development for the gasification of biomass [151].

8.6 Conclusions

Esterification and transesterification are the two main processes in the manufacture of biodiesel. The type of feedstock oil, the catalyst utilized, the reaction conditions, and the molar ratio of alcohol to oil all have an impact on these reactions. By removing issues associated with the commonly used reaction schemes (which depend on homogeneous catalysts, typically H_2SO_4, KOH, or NaOH), the use of carbon-based catalysts in these reactions opens the door to cost reduction and environmentally friendly biodiesel production. According to many studies, carbon can be used as a substrate for a range of active catalysts (metals, metal oxides, and so on) because of its exceptional surface and structural characteristics, as well as its great thermal stability [39]. By attaching different acidic or basic functional groups to activated carbon (AC), which has a structural resemblance to graphite or graphene sheets, researchers can create new materials. Such materials have distinct structural features based on the carbon materials and catalytic properties that depend on the nature of the attached molecule/group (acidic or basic). These materials may be used as heterogeneous catalysts in a variety of reactions, including the synthesis of biodiesel [42].

8.6.1 Environmental impact and sustainability of carbon-based materials

To ensure sustainability, it is essential to weigh the advantages of carbon-based products against their effects on the environment. Sustainable practices, such as developing biodegradable alternatives using carbon-based materials from renewable resources like cornstarch or sugarcane, can reduce reliance on fossil fuels and minimize environmental impact. These materials decompose more readily, thereby reducing pollution. Carbon-based materials serve as economical, metal-free substitutes and offer environmentally friendly solutions for energy conversion and storage [152]. These materials are considered promising advanced materials because they are abundant, environmentally friendly, structurally tunable at the atomic/morphological levels, and mechanically robust. They also offer a large and modifiable surface area, adjustable porosity and form, and remarkable stability against harsh conditions, regeneration, and undesired reactions. In recent decades, activated carbons have been investigated for use in several heterogeneous catalysis processes [153–155].

Among the many benefits of carbon nanomaterials are their excellent surface properties, small size, biocompatibility (for the most part), simplicity in synthesis, and obvious application in a variety of industries. However, the focus today is on researching the potentially harmful impacts of nanoparticles on living things. There is reason for caution, as it is necessary to thoroughly investigate the long-term consequences of the widespread use of nanoparticles and their release into the environment [45]. Implementing carbon-based material recycling schemes can also greatly reduce pollution, greenhouse gas emissions, and environmental effects. Furthermore, the sustainability of carbon-based materials can be improved by developments in production technology. Additionally, the carbon footprint related to material production can be reduced by incorporating renewable energy sources into manufacturing operations.

LCAs are necessary to understand the environmental impacts of carbon-based materials. LCAs evaluate the total environmental impact from raw material extraction to disposal, allowing for informed material selection and usage decisions. Sustainable design and production methods are encouraged by this all-encompassing strategy [155]. With more research and interdisciplinary collaboration in process optimization, reactor design, green route development, and molecular simulation, it is expected that the feasibility and environmental friendliness of heterogeneous catalyst-based biorefinery processes will increase, supporting sustainable development [156].

8.6.2 Challenges, research gaps, and opportunities

Carbon-based catalysts show enhanced potential to produce biofuel due to their accessibility, affordability, flexibility to change hydrophobicity and polarity, durability at high temperatures and in a range of chemical conditions, and ease of extraction from reaction mixtures [45]. A highly effective heterogeneous catalyst should have favorable pore size, surface area, and active site properties, together with a low risk of leaching, in order to achieve optimal catalytic performance. Although earlier research has shown promising results, issues that arise during catalyst preparation can have a direct impact on a catalyst's efficacy and can differ according to the technique employed. In some cases, the reaction rate may be slow, and leaching may also occur. In addition to being economically viable, a good solid catalyst should be nontoxic, noncorrosive, easily separated, readily available, and safe to handle. The high temperatures required to create catalysts, however, increase costs and energy consumption. Additionally, it has been observed that certain methods for producing a catalyst from agricultural waste increase costs and demands on the catalyst synthesis process [157]. More attention must be paid to preparation methods for transforming solid waste into catalysts; such processes should be affordable, easy to use, and environmentally benign. High-temperature carbonization or activation is a crucial recycling phase, as it uses a lot of energy and produces secondary environmental contamination.

The efficiency and feasibility of biodiesel synthesis ultimately depend on the stability of the catalysts used. Therefore, it is suggested that future research should focus on efficient ways of producing catalysts that are stable in the long term, as this would increase the likelihood of beneficial commercial applications. Green biofuel production methods must be adopted as the world shifts its focus to a sustainable and renewable economy. Significant progress in the use of heterogeneous catalysts in biomass-to-fuel production is still required to achieve efficient and cost-effective processes. New multifunctional chemicals, single-step reaction techniques, immobilized catalysts, and nanocatalysts should be the focus of future research. Priority should also be given to the development of easier and less expensive catalyst preparation methods. To improve catalyst stability, alternative catalyst types with superior compositional and structural tailorability, such as metal–organic frameworks, encapsulated nanocatalysts, and sandwich-structured nanocatalysts, should be developed [157].

In summary, significant progress has been made in developing heterogeneous catalysts from bioderived precursors for biodiesel production, although certain problems with their synthesis remain. These include the consistent dispersion of active ingredients, thermal stability, scalability, effects on the environment and the economy, and reproducibility. Achieving uniform and consistent active phase dispersion is also a significant problem in heterogeneous catalyst production. Several problems need to be addressed before these catalysts can be used in industrial biodiesel synthesis. Despite the advantages of these catalysts, heterogeneous catalyst recovery remains challenging, particularly when separation techniques such as centrifugation and filtration are employed, which can result in significant catalyst losses. More research is needed to develop increasingly advanced and effective catalytic conversion methods [156].

References and further reading

[1] Novoselov K S, Geim A K, Morozov S V, Jiang D, Zhang Y, Dubonos S V, Grigorieva I V and Firsov A A 2004 Electric field in atomically thin carbon films *Science* **306** 666–9

[2] Ahmad A L, Jawad Z A, Low S C and Zein S H S 2014 A cellulose acetate/multi-walled carbon nanotube mixed matrix membrane for CO_2/N_2 separation *J. Memb. Sci.* **451** 55–66

[3] Iijima S 1991 Helical microtubules of graphitic carbon *Nature* **354** 56–8

[4] IEA 2019 EIA: international energy outlook 2019 presentation *Int. Energy Agency* **2019** 18 www.eia.gov/ieo

[5] IEA 2023 *World Energy Outlook 2023* (Paris: IEA) https://iea.org/reports/world-energy-outlook-2023, Licence: CC BY 4.0 (report); CC BY NC SA 4.0 (Annex A)

[6] Dunn B, Kamath H and Tarascon J-M 2011 Electrical energy storage for the grid: a battery of choices *Science* **334** 928–35

[7] Hasani A, Teklagne M A, Do H H, Hong S H, Van Le Q, Ahn S H and Kim S Y 2020 Graphene-based catalysts for electrochemical carbon dioxide reduction *Carbon Energy* **2** 158–75

[8] Dresselhaus M S and Avouris P 2001 Introduction to carbon materials research *BT— Carbon Nanotubes: Synthesis, Structure, Properties, and Applications* ed M S Dresselhaus, G Dresselhaus and P Avouris (Berlin: Springer) 1–9

[9] Dresselhaus M S and Endo M 2001 Relation of carbon nanotubes to other carbon materials *BT—Carbon Nanotubes: Synthesis, Structure, Properties, and Applications* ed M S Dresselhaus, G Dresselhaus and P Avouris (Berlin: Springer) 11–28

[10] Zhang L L and Zhao X S 2009 Carbon-based materials as supercapacitor electrodes *Chem. Soc. Rev.* **38** 2520–31

[11] Chu S and Majumdar A 2012 Opportunities and challenges for a sustainable energy future *Nature* **488** 294–303

[12] Tarascon J M and Armand M 2001 Issues and challenges facing rechargeable lithium batteries *Nature* **414** 359–67

[13] Simon P and Gogotsi Y 2008 Materials for electrochemical capacitors *Nat. Mater.* **7** 845–54

[14] Sevilla M and Mokaya R 2014 Energy storage applications of activated carbons: super-capacitors and hydrogen storage *Energy Environ. Sci.* **7** 1250–80

[15] Wang H, Yang Y, Liang Y, Robinson J T, Li Y, Jackson A, Cui Y and Dai H 2011 Graphene-wrapped sulfur particles as a rechargeable lithium-sulfur battery cathode material with high capacity and cycling stability *Nano Lett.* **11** 2644–7

[16] Wang Q, Hu W and Huang Y 2017 Nitrogen doped graphene anchored cobalt oxides efficiently bi-functionally catalyze both oxygen reduction reaction and oxygen revolution reaction *Int. J. Hydrogen Energy.* **42** 5899–907

[17] Qu L, Liu Y, Baek J-B and Dai L 2010 Nitrogen-doped graphene as efficient metal-free electrocatalyst for oxygen reduction in fuel cells *ACS Nano* **4** 1321–6

[18] Marsh H and Rodríguez-Reinoso F 2006 Activated carbon (origins) *Activated Carbon* ed H Marsh and F Rodríguez-Reinoso (Oxford: Elsevier Science Ltd) 13–86 ch 2

[19] Adegoke K A and Maxakato N W 2022 Efficient strategies for boosting the performance of 2D graphitic carbon nitride nanomaterials during photoreduction of carbon dioxide to energy-rich chemicals *Mater. Today Chem.* **23** 100605

[20] Bonaccorso F, Colombo L, Yu G, Stoller M, Tozzini V, Ferrari A C, Ruoff R S and Pellegrini V 2015 Graphene, related two-dimensional crystals, and hybrid systems for energy conversion and storage *Science* **347** 1246501

[21] Lefèvre M, Proietti E, Jaouen F and Dodelet J P 2009 Iron-based catalysts with improved oxygen reduction activity in polymer electrolyte fuel cells *Science* **324** 71–4

[22] Adegoke K A, Oyebamiji A K, Adeola A O, Olabintan A B, Oyedotun K O, Mamba B B and Bello O S 2024 Iron-based metal–organic frameworks and derivatives for electro-chemical energy storage and conversion *Coord. Chem. Rev.* **517** 215959

[23] Logan B E and Rabaey K 2012 Conversion of wastes into bioelectricity and chemicals by using microbial electrochemical technologies *Science* **337** 686–90

[24] Yu C, Kim Y S, Kim D and Grunlan J C 2008 Thermoelectric behavior of segregated-network polymer nanocomposites *Nano Lett.* **8** 4428–32

[25] Wang G, Zhang L and Zhang J 2012 A review of electrode materials for electrochemical supercapacitors *Chem. Soc. Rev.* **41** 797–828

[26] Frackowiak E and Béguin F 2001 Carbon materials for the electrochemical storage of energy in capacitors *Carbon N. Y.* **39** 937–50

[27] Ji X and Nazar L F 2010 Advances in Li–S batteries *J. Mater. Chem.* **20** 9821–6

[28] Dreyer D R, Park S, Bielawski C W and Ruoff R S 2010 The chemistry of graphene oxide *Chem. Soc. Rev.* **39** 228–40

[29] Cossutta M, Vretenar V, Centeno T A, Kotrusz P, McKechnie J and Pickering S J 2020 A comparative life cycle assessment of graphene and activated carbon in a supercapacitor application *J. Clean. Prod.* **242** 118468

[30] Balat M 2010 Potential alternatives to edible oils for biodiesel production—a review of current work *Energy Convers. Manag.* **52** 1479–92

[31] Priya A, Hu Y, Mou J, Du C, Wilson K, Luque R and Sze C 2023 Introduction: an overview of biofuels and production technologies *Elsevier EBooks* 3–24

[32] Koçar G and Civaş N 2013 An overview of biofuels from energy crops: current status and future prospects *Renew. Sustain. Energy Rev.* **28** 900–16

[33] Agarwal A K, Agarwal R A, Gupta T and Gurjar B R 2017 Introduction to biofuels *Biofuels: Green Energy and Technology* (Singapore: Springer) 3–6

[34] Arshad M, Zia M A, Shah F A and Ahmad M 2017 *An Overview of Biofuel* (Cham: Springer) 1–37

[35] Karwat D M A, Eagle W E and Wooldridge M S 2014 Are there ecological problems that technology cannot solve? Water scarcity and dams, climate change and biofuels *Int. J. Eng. Soc. Just. Peace.* **3** 7–25

[36] Balat M 2011 Potential alternatives to edible oils for biodiesel production—a review of current work *Energy Convers. Manag.* **52** 1479–92

[37] Balat M 2011 Production of bioethanol from lignocellulosic materials via the biochemical pathway: a review *Energy Convers. Manag.* **52** 858–75

[38] Jeswani H K, Chilvers A and Azapagic A 2020 Environmental sustainability of biofuels: a review *Proc. R. Soc. A Math. Phys. Eng. Sci.* **476** 20200351

[39] Konwar L J, Boro J and Deka D 2014 Review on latest developments in biodiesel production using carbon-based catalysts *Renew. Sustain. Energy Rev.* **29** 546–64

[40] Melero J A, Iglesias J and Morales G 2009 Heterogeneous acid catalysts for biodiesel production: current status and future challenges *Green Chem.* **11** 1285

[41] Ejikeme P M, Anyaogu I D, Ejikeme C L, Nwafor N P, Egbuonu C A C, Ukogu K and Ibemesi J A 2009 Catalysis in biodiesel production by transesterification processes: an insight *J. Chem.* **7** 1120–32

[42] Ballotin F C, Lago R M and Paula A 2020 Solid acid catalysts based on sulfonated carbon nanostructures embedded in an amorphous matrix produced from bio-oil: esterification of oleic acid with methanol *J. Environ. Chem. Eng.* **8** 103674

[43] Clohessy J and Kwapinski W 2020 Carbon-based catalysts for biodiesel production—a review *Appl. Sci.* **10** 918

[44] Hara M 2009 Environmentally benign production of biodiesel using heterogeneous catalysts *ChemSusChem.* **2** 129–35

[45] Khan N R and Bin Rashid A 2024 Carbon-based nanomaterials: a paradigm shift in biofuel synthesis and processing for a sustainable energy future *Energy Convers. Manag. X.* **22** 100590

[46] Bhagat R, Panakkal H, Gupta I and Ingle A P 2021 Carbon-based nanocatalysts in biodiesel production *Nano- and Biocatalysts for Biodiesel Production* ed A P Ingle (New York, NY: Wiley) 157–81

[47] Xia D, Yu H, Li H, Huang P, Li Q and Wang Y 2022 Carbon-based and carbon-supported nanomaterials for the catalytic conversion of biomass: a review *Environ. Chem. Lett.* **20** 1719–44

[48] Mo X, Lotero E, Lu C, Liu Y and Goodwin J G 2008 A novel sulfonated carbon composite solid acid catalyst for biodiesel synthesis *Catal. Lett.* **123** 1–6

[49] Rao B V S, Mouli K C, Rambabu N, Dalai A K and Prasad R B N 2011 Carbon-based solid acid catalyst from de-oiled canola meal for biodiesel production *Catal. Commun.* **14** 20–6

[50] Shu Q, Nawaz Z, Gao J, Liao Y, Zhang Q, Wang D and Wang J 2010 Synthesis of biodiesel from a model waste oil feedstock using a carbon-based solid acid catalyst: reaction and separation *Bioresour. Technol.* **101** 5374–84

[51] Liu R, Wang X, Zhao X and Feng P 2008 Sulfonated ordered mesoporous carbon for catalytic preparation of biodiesel *Carbon N. Y.* **46** 1664–9

[52] Yuan C, Abomohra A E-F, Wang S, Liu Q, Zhao S, Cao B, Hu X, Marrakchi F, He Z and Hu Y 2021 High-grade biofuel production from catalytic pyrolysis of waste clay oil using modified activated seaweed carbon-based catalyst *J. Clean. Prod.* **313** 127928

[53] Liao W, Zhu Z, Chen N, Su T, Deng C, Zhao Y, Ren W and Lü H 2019 Highly active bifunctional Pd-Co9S8/S-CNT catalysts for selective hydrogenolysis of 5-hydroxymethyl-furfural to 2,5-dimethylfuran *Mol. Catal.* **482** 110756

[54] Pua F, Fang Z, Zakaria S, Guo F and Chia C 2011 Direct production of biodiesel from high-acid value Jatrophaoil with solid acid catalyst derived from lignin *Biotechnol. Biofuels* **4** 56

[55] Guo F, Xiu Z-L and Liang Z-X 2012 Synthesis of biodiesel from acidified soybean soapstock using a lignin-derived carbonaceous catalyst *Appl. Energy* **98** 47–52

[56] Geng L, Yu G, Wang Y and Zhu Y 2012 Ph-SO3H-modified mesoporous carbon as an efficient catalyst for the esterification of oleic acid *Appl. Catal. A Gen.* 427–8 137–44

[57] Macina A, de V and Naccache R 2019 A carbon dot-catalyzed transesterification reaction for the production of biodiesel *J. Mater. Chem.* A **7** 23794–802

[58] Rentería J, Gallego A, Gamboa D, Cacua K and Herrera B 2022 Effect of amide-functionalized carbon nanotubes as commercial diesel and palm-oil biodiesel additives on the ignition delay: a study on droplet scale *Fuel* **338** 127202

[59] Hara M 2010 Biodiesel production by amorphous carbon bearing SO₃H, COOH and phenolic OH groups, a solid brønsted acid catalyst *Top. Catal.* **53** 805–10

[60] Toda M, Takagaki A, Okamura M, Kondo J N, Hayashi S, Domen K and Hara M 2005 Biodiesel made with sugar catalyst *Nature* **438** 178

[61] Villa A, Tessonnier J-P, Majoulet O, Su D S and Schlögl R 2009 Amino-functionalized carbon nanotubes as solid basic catalysts for the transesterification of triglycerides *Chem. Commun.* **2009** 4405–7

[62] Villa A, Tessonnier J, Majoulet O, Su D S and Schlögl R 2010 Transesterification of triglycerides using nitrogen-functionalized carbon nanotubes *ChemSusChem.* **3** 241–5

[63] Yuan C, Chen W and Yan L 2012 Amino-grafted graphene as a stable and metal-free solid basic catalyst *J. Mater. Chem.* **22** 7456

[64] Xu X, Li Y, Gong Y, Zhang P, Li H and Wang Y 2012 Synthesis of palladium nanoparticles supported on mesoporous N-doped carbon and their catalytic ability for biofuel upgrade *J. Am. Chem. Soc.* **134** 16987–90

[65] Hasannia S, Kazemeini M and Seif A 2024 Optimizing parameters for enhanced rapeseed biodiesel production: a study on acidic and basic carbon-based catalysts through experimental and DFT evaluations *Energy Convers. Manag.* **303** 118201

[66] Mardhiah H H, Ong H C, Masjuki H H, Lim S and Pang Y L 2017 Investigation of carbon-based solid acid catalyst from Jatropha curcas biomass in biodiesel production *Energy Convers. Manag.* **144** 10–7

[67] Bastos R R C, da Luz Corrêa A P, da Luz P T S, da Rocha Filho G N, Zamian J R and da Conceição L R V 2020 Optimization of biodiesel production using sulfonated carbon-based catalyst from an Amazon agro-industrial waste *Energy Convers. Manag.* **205** 112457

[68] Rana A, Alghazal M S M, Alsaeedi M M, S. Bakdash R, Basheer C and Al-Saadi A A 2019 Preparation and characterization of biomass carbon–based solid acid catalysts for the esterification of marine algae for biodiesel production *BioEnergy Res.* **12** 433–42

[69] Roy M and Mohanty K 2021 Valorization of de-oiled microalgal biomass as a carbon-based heterogeneous catalyst for a sustainable biodiesel production *Bioresour. Technol.* **337** 125424

[70] Tang Z-E, Lim S, Pang Y-L, Shuit S-H and Ong H-C 2020 Utilisation of biomass wastes based activated carbon supported heterogeneous acid catalyst for biodiesel production *Renew. Energy.* **158** 91–102

[71] Tran H Q and Dinh N T 2018 Study on the preparation of ordered mesoporous carbon-based catalyst from waste microalgal biomass for the synthesis of biokerosene *J. Porous Mater.* **25** 1567–76

[72] Dhawane S H, Kumar T and Halder G 2016 Biodiesel synthesis from Hevea brasiliensis oil employing carbon-supported heterogeneous catalyst: optimization by Taguchi method *Renew. Energy.* **89** 506–14

[73] Ruatpuia J V L, Changmai B, Pathak A, Alghamdi L A, Kress T, Halder G, Wheatley A E H and Rokhum S L 2023 Green biodiesel production from Jatropha curcas oil using a carbon-based solid acid catalyst: a process optimization study *Renew. Energy.* **206** 597–608

[74] Zhang M, Sun A, Meng Y, Wang L, Jiang H and Li G 2014 Catalytic performance of biomass carbon-based solid acid catalyst for esterification of free fatty acids in waste cooking oil *Catal. Surv. Asia.* **19** 61–7

[75] Roberto R, Paula A, Souza T, Narciso G, Zamian J R and Vieira R 2020 Optimization of biodiesel production using sulfonated carbon-based catalyst from an Amazon agro-industrial waste *Energy Convers. Manag.* **205** 112457

[76] Alper E and Yuksel Orhan O 2017 CO_2 utilization: developments in conversion processes *Petroleum* **3** 109–26

[77] Yusuf N, Almomani F and Qiblawey H 2023 Catalytic CO_2 conversion to C1 value-added products: review on latest catalytic and process developments *Fuel* **345** 128178

[78] Whang H S, Lim J, Choi M S, Lee J and Lee H 2019 Heterogeneous catalysts for catalytic CO_2 conversion into value-added chemicals *BMC Chem. Eng.* **1** 1–19

[79] Duan X, Xu J, Wei Z, Ma J, Guo S, Wang S, Liu H and Dou S 2017 Metal-free carbon materials for CO_2 electrochemical reduction *Adv. Mater.* **29** 1701784

[80] Aggarwal M, Basu S, Shetti N P, Nadagouda M N, Kwon E E, Park Y-K and Aminabhavi T M 2021 Photocatalytic carbon dioxide reduction: exploring the role of ultrathin 2D graphitic carbon nitride (g-C_3N_4) *Chem. Eng. J.* **425** 131402

[81] Garcia J A, Villen-Guzman M, Rodriguez-Maroto J M and Paz-Garcia J M 2022 Technical analysis of CO_2 capture pathways and technologies *J. Environ. Chem. Eng.* **10** 108470

[82] Wu Q-J, Liang J, Huang Y-B and Cao R 2022 Thermo-, electro-, and photocatalytic CO_2 conversion to value-added products over porous metal/covalent organic frameworks *Acc. Chem. Res.* **55** 2978–97

[83] Debbie Y, Joo J, Wu W-Y, Tao L, Wang C, Zhu Q and Bu J 2024 Advancements in CO_2 capture by absorption and adsorption: a comprehensive review *J. CO2 Util.* **81** 102727

[84] Dissanayake P D, Choi S W, Igalavithana A D, Yang X, Tsang D C W, Wang C-H, Kua H W, Lee K B and Ok Y S 2020 Sustainable gasification biochar as a high efficiency adsorbent for CO_2 capture: a facile method to designer biochar fabrication *Renew. Sustain. Energy Rev.* **124** 109785

[85] Zhao K and Quan X 2021 Carbon-based materials for electrochemical reduction of CO_2 to C^{2+} oxygenates: recent progress and remaining challenges *ACS Catal.* **11** 2076–97

[86] Sundar D, Liu C-H, Anandan S and Wu J J 2023 Photocatalytic CO_2 conversion into solar fuels using carbon-based materials—a review *Molecules* **28** 5383

[87] Shao J, Ma C, Zhao J, Wang L and Hu X 2021 Effective nitrogen and sulfur co-doped porous carbonaceous CO_2 adsorbents derived from amino acid *Colloids Surf. A: Physicochem. Eng. Asp.* **632** 127750

[88] Zhang T *et al* 2021 Regulation of functional groups on graphene quantum dots directs selective CO_2 to CH_4 conversion *Nat. Commun.* **12** 5265

[89] Yue Y, Sun Y, Tang C, Liu B, Ji Z, Hu A, Shen B, Zhang Z and Sun Z 2019 Ranking the relative CO_2 electrochemical reduction activity in carbon materials *Carbon N. Y.* **154** 108–14

[90] Ye S, Fan G, Xu J, Yang L and Li F 2020 Nickel-nitrogen-modified porous carbon/carbon nanotube hybrid with necklace-like geometry: an efficient and durable electrocatalyst for selective reduction of CO_2 to CO in a wide negative potential region *Electrochim. Acta* **334** 135583

[91] Razzaq A, Ali S, Asif M and In S-I 2020 Layered double hydroxide (LDH) based photocatalysts: an outstanding strategy for efficient photocatalytic CO_2 conversion *Catalysts* **10** 1185

[92] Kandy M M 2019 Carbon-based photocatalysts for enhanced photocatalytic reduction of CO_2 to solar fuels *Sustain. Energy Fuels.* **4** 469–84

[93] Xu M, Jiang H, Li X, Gao M, Liu Q, Wang H, Huo P and Chen S 2021 Design of a ZnS/CdS/rGO composite nanosheet photocatalyst with multi-interface electron transfer for high conversion of CO_2 *Sustain. Energy Fuels.* **5** 4606–17

[94] Crake A, Christoforidis K C, Godin R, Moss B, Kafizas A, Zafeiratos S, Durrant J R and Petit C 2019 Titanium dioxide/carbon nitride nanosheet nanocomposites for gas phase CO_2 photoreduction under UV–visible irradiation *Appl. Catal. B Environ.* **242** 369–78

[95] Li X, Jiang H, Ma C, Zhu Z, Song X, Wang H, Huo P and Li X 2021 Local surface plasma resonance effect enhanced Z-scheme ZnO/Au/g-C_3N_4 film photocatalyst for reduction of CO_2 to CO *Appl. Catal. B Environ.* **283** 119638

[96] Bharath G, Prakash J, Rambabu K, Venkatasubbu G D, Kumar A, Lee S, Theerthagiri J, Choi M Y and Banat F 2021 Synthesis of TiO_2/RGO with plasmonic Ag nanoparticles for highly efficient photoelectrocatalytic reduction of CO_2 to methanol toward the removal of an organic pollutant from the atmosphere *Environ. Pollut.* **281** 116990

[97] Li Z, Liu Z, Li Y and Wang Q 2021 Flower-like CoAl layered double hydroxides modified with CeO_2 and RGO as efficient photocatalyst towards CO_2 reduction *J. Alloys Compd.* **881** 160650

[98] Li X, Jiang H, Ma C, Zhu Z, Song X, Wang H, Huo P and Li X 2020 Local surface plasma resonance effect enhanced Z-scheme ZnO/Au/g-C_3N_4 film photocatalyst for reduction of CO_2 to CO *Appl. Catal. B Environ. Energy* **283** 119638

[99] Wang H, Liu Q, Xu M, Yan C, Song X, Liu X, Wang H, Zhou W and Huo P 2023 Dual-plasma enhanced 2D/2D/2D g-C_3N_4/Pd/MoO_{3-x} S-scheme heterojunction for high-selectivity photocatalytic CO_2 reduction *Appl. Surf. Sci.* **640** 158420

[100] Fu Z *et al* 2020 A stable covalent organic framework for photocatalytic carbon dioxide reduction *Chem. Sci.* **11** 543–50

[101] Ye L, Deng Y, Wang L, Xie H and Su F 2019 Bismuth-based photocatalysts for solar photocatalytic carbon dioxide conversion *ChemSusChem.* **12** 3671–701

[102] Adenuga D O, Tichapondwa S M and Chirwa E M N 2020 Facile synthesis of a Ag/AgCl/BiOCl composite photocatalyst for visible-light-driven pollutant removal *J. Photochem. Photobiol. A: Chem.* **401** 112747

[103] Rodríguez V, Camarillo R, Martínez F, Jiménez C and Rincón J 2021 High-pressure synthesis of rGO/TiO$_2$ and rGO/TiO$_2$/Cu catalysts for efficient CO$_2$ reduction under solar light *J. Supercrit. Fluids* **174** 105265

[104] Kandy M M and Gaikar V G 2019 Continuous photocatalytic reduction of CO$_2$ using nanoporous reduced graphene oxide (RGO)/cadmium sulfide (CdS) as catalyst on porous anodic alumina (PAA)/aluminum support *J. Nanosci. Nanotechnol.* **19** 5323–31

[105] Li Y, Li X, Zhang H, Fan J and Xiang Q 2020 Design and application of active sites in g-C3N4-based photocatalysts *J. Mater. Sci. Technol.* **56** 69–88

[106] Prabhu P, Jose V and Lee J 2020 Heterostructured catalysts for electrocatalytic and photocatalytic carbon dioxide reduction *Adv. Funct. Mater.* **30** 1910768

[107] He Y, Zhang L, Teng B and Fan M 2014 New application of Z-scheme Ag$_3$PO$_4$/g-C$_3$N$_4$ composite in converting CO$_2$ to fuel *Environ. Sci. Technol.* **49** 649–56

[108] Adenuga D, Skosana S, Tichapondwa S and Chirwa E 2021 Synthesis of a plasmonic AgCl and oxygen-rich Bi$_{24}$O$_{31}$Cl$_{10}$ composite heterogeneous catalyst for enhanced degradation of tetracycline and 2,4-dichlorophenoxy acetic acid *RSC Adv.* **11** 36760–8

[109] Lu M, Li Q, Zhang C, Fan X, Li L, Dong Y, Chen G and Shi H 2020 Remarkable photocatalytic activity enhancement of CO$_2$ conversion over 2D/2D g-C$_3$N$_4$/BiVO$_4$ Z-scheme heterojunction promoted by efficient interfacial charge transfer *Carbon N. Y.* **160** 342–52

[110] Bakhoum D T, Oyedotun K O, Sarr S, Sylla N F, Maphiri V M, Ndiaye N M, Ngom B D and Manyala N 2022 A study of porous carbon structures derived from composite of cross-linked polymers and reduced graphene oxide for supercapacitor applications *J. Energy Storage.* **51** 104476

[111] Rabbani S, Qureshi Z A, Alfantazi A, Alkaabi A K, Alameri S A, Addad Y, Samad Y A and Afgan I 2024 Graphene-nuclear nexus: a critical review *2D Mater.* **11** 42001

[112] Li M, Chen T, Gooding J J and Liu J 2019 Review of carbon and graphene quantum dots for sensing *ACS Sens.* **4** 1732–48

[113] Khan J, Momin S A and Mariatti M 2020 A review on advanced carbon-based thermal interface materials for electronic devices *Carbon N. Y.* **168** 65–112

[114] Gao Q *et al* 2023 Commercial carbon nanotube as rear contacts for industrial p-type silicon solar cells with an efficiency exceeding 23% *Carbon N. Y.* **202** 432–7

[115] Chen M, Wang N, Bai D, Li Y, Yang S, Wang Z, Zhu X, Yang D and Liu S F 2025 Revolutionizing perovskite solar cells with carbon electrodes: innovations and economic potential *Adv. Funct. Mater.* **35** 2422020

[116] Islam M N, Chowdhury M R H, Chowdhury M H D, Islam M J, Rahaman M L and Chowdhury M I B 2024 A high-efficiency dye-sensitized solar cell using PCDTBT:PCBM as solid electrolyte and graphene oxide as hole transport layer *2024 6th Int. Conf. Electr. Eng. Inf. Commun. Technol.* 933–8

[117] Upadhyay S, Narendhiran S and Balachandran M 2025 Cotton-derived carbon fibers and MoS$_2$ hybrids for efficient I^{3-} reduction in bifacial dye-sensitized solar cells *Carbon N. Y.* **238** 120248

[118] Xiao P *et al* 2025 13.7% efficient graphene/Si heterojunction solar cells with one-step transferred polymer anti-reflection layer for enhanced light absorption and device durability *Chem. Asian J.* **20** e202401816

[119] Sharma R K, Srivastava A, Kumar A, Prajapat P, Tawale J S, Pathi P, Gupta G and Srivastava S K 2024 Graphene oxide as an effective interface passivation layer for enhanced performance of hybrid silicon solar cells *ACS Appl. Energy Mater.* **7** 4710–24

[120] Yan J *et al* 2021 Stable organic passivated carbon nanotube–silicon solar cells with an efficiency of 22% *Adv. Sci.* **8** 2102027

[121] Maurel G, Nya F T and Laref A 2024 Organic solar cell efficiency improvement through architecture engineering by integrating carbon nanoring—SCAPS 1D modelling *Mater. Chem. Phys.* **324** 129713

[122] Nguyen D C T, Kim B-S, Oh G-H, Vu V-P, Kim S and Lee S-H 2023 Incorporation of carbon quantum dots with PEDOT:PSS for high-performance inverted organic solar cells *Synth. Met.* **298** 117430

[123] Fang H *et al* 2024 Fullerene-hybridized fused-ring electron acceptor with high dielectric constant and isotropic charge transport for organic solar cells *Angew. Chemie Int. Ed.* **64** e202417951

[124] Kim H-S *et al* 2012 Lead iodide perovskite sensitized all-solid-state submicron thin film mesoscopic solar cell with efficiency exceeding 9% *Sci. Rep.* **2** 591

[125] Jiang Q *et al* 2022 Surface reaction for efficient and stable inverted perovskite solar cells *Nature* **611** 278–83

[126] Ku Z, Rong Y, Xu M, Liu T and Han H 2013 Full printable processed mesoscopic $CH_3NH_3PbI_3/TiO_2$ heterojunction solar cells with carbon counter electrode *Sci. Rep.* **3** 3132

[127] Elseman A M and Al-Gamal A G 2024 Multiwalled carbon nanotubes as hole collectors in inverted perovskite solar cells *ACS Appl. Nano Mater.* **7** 8792–803

[128] Qiu L, Tian W, Xu M, Xiao J, Liang J, Liu F and Zhao Y 2025 Surface reinforcement of perovskite films with heteroatom-modulated carbon nanosheets for heat-resistant solar cells *Dalt. Trans.* **54** 8204–13

[129] Cipriani G, Boscaino V, Di Dio V, Cardona F, Zizzo G, Di Caro S and Sa'ed J 2019 Application of thermographic techniques for the detection of failures on photovoltaic modules *2019 IEEE Int. Conf. Environ. Electr. Eng. 2019 IEEE Ind. Commer. Power Syst. Eur. (EEEIC/I&CPS Eur.)* 1–5

[130] Moh T S Y, Ting T W and Lau A H Y 2020 Graphene Nanoparticles (GNP) nanofluids as key cooling media on a flat solar panel through micro-sized channels *Energy Rep.* **6** 282–6

[131] Kargaran M, Goshayeshi H R, Azarberahman S and Chaer I 2025 Advanced cooling of photovoltaic panels using hybrid nanofluids incorporating graphene oxide and carbon nanotubes *Int. J. Energy Res.* **2025** 4345236

[132] Tuncer A D, Badali Y and Khanlari A 2024 Application of carbon-based nanomaterials in solar-thermal systems: an updated review *Sol. Energy* **282** 112932

[133] Ren H *et al* 2017 Hierarchical graphene foam for efficient omnidirectional solar–thermal energy conversion *Adv. Mater.* **29** 1702590

[134] Sattar A, Bofeng B, Munir M A, Farooq M, Bilal S, Khan M I, Akbar N S, Khan M I, Rehan M and Riaz F 2025 Photothermal performance of nanofluids: an experimental study on direct absorption solar energy conversion using graphene oxide and its binary composites for water purification *Energy Convers. Manag. X* **26** 100898

[135] Chen Y, Zhang Q, Wen X, Yin H and Liu J 2018 A novel CNT encapsulated phase change material with enhanced thermal conductivity and photo-thermal conversion performance *Sol. Energy Mater. Sol. Cells* **184** 82–90

[136] Bai Y, Lin F, Liu X, Feng J, Zhu X, Huang Z, Min X, Mi R and Qiao J 2022 Directional chitosan/carbon fiber powder aerogel supported phase change composites for effective solar thermal energy conversion and hot compression *J. Energy Storage.* **56** 105980

[137] Moulefera I, Marín J J D, Cascales A, Montalbán M G, Alarcón M and Víllora G 2025 Innovative application of graphene nanoplatelet-based ionanofluids as heat transfer fluid in hybrid photovoltaic-thermal solar collectors *Sci. Rep.* **15** 6489

[138] Kazemian A, Salari A, Ma T and Lu H 2022 Application of hybrid nanofluids in a novel combined photovoltaic/thermal and solar collector system *Sol. Energy* **239** 102–16

[139] Megía P J, Vizcaíno A J, Calles J A and Carrero A 2021 Hydrogen production technologies: from fossil fuels toward renewable sources. A mini review *Energy Fuels* **35** 16403–15

[140] Bak T, Nowotny J, Rekas M and Sorrell C C 2002 Photo-electrochemical hydrogen generation from water using solar energy. Materials-related aspects *Int. J. Hydrogen Energy* **27** 991–1022

[141] Tee S Y, Win K Y, Teo W S, Koh L, Liu S, Teng C P and Han M 2017 Recent progress in energy-driven water splitting *Adv. Sci.* **4** 1600337

[142] Yan Y, Zhai D, Liu Y, Gong J, Chen J, Zan P, Zeng Z, Li S, Huang W and Chen P 2020 van der Waals heterojunction between a bottom-up grown doped graphene quantum dot and graphene for photoelectrochemical water splitting *ACS Nano* **14** 1185–95

[143] Peng G, Albero J, Garcia H and Shalom M 2018 A water-splitting carbon nitride photoelectrochemical cell with efficient charge separation and remarkably low onset potential *Angew. Chem. Int. Ed.* **57** 15807–11

[144] Alenad A M, Taha T A, Zayed M, Gamal A, Shaaban M, Ahmed A M and Mohamed F 2023 Impact of carbon nanotubes concentrations on the performance of carbon nanotubes/zinc oxide nanocomposite for photoelectrochemical water splitting *J. Electroanal. Chem.* **943** 117579

[145] Chaulagain N *et al* 2025 Heteroepitaxial growth of narrow band gap carbon-rich carbon nitride using *in situ* polymerization to empower sunlightdriven photoelectrochemical water splitting *J. Am. Chem. Soc.* **147** 11511–32

[146] Elbakkay M H, El Rouby W M A, El-Dek S I and Farghali A A 2022 S-TiO$_2$/S-reduced graphene oxide for enhanced photoelectrochemical water splitting *Appl. Surf. Sci.* **439** 1088–102

[147] Cao L, Yu I K M, Xiong X, Tsang D C W, Zhang S, Clark J H, Hu C, Ng Y H, Shang J and Ok Y S 2020 Biorenewable hydrogen production through biomass gasification: a review and future prospects *Environ. Res.* **186** 109547

[148] Li G, Yang Y, Yu Q, Ma Q, Lam S S, Chen X, He Y, Ge S, Sonne C and Peng W 2024 Application of nanotechnology in hydrogen production from biomass: a critical review *Adv. Compos. Hybrid Mater.* **7** 17

[149] Hussain S, Irmak S and Farid M U 2025 Developing N,S-doped hierarchical porous carbon-supported Pt catalysts for hydrothermal gasification of woody biomass to hydrogen *Next Energy* **8** 100257

[150] Bamaca E, Irmak S, Wilkins M and Smith T 2021 Effect of precursors on graphene-supported platinum monometalic catalysts for hydrothermal gasification of biomass compounds to hydrogen *Fuel* **290** 120079

[151] Ali M A M, Inoue S and Matsumura Y 2022 Carbon nanotube as catalyst support in supercritical water *J. Supercrit. Fluids* **190** 105755

[152] Liu X and Dai L 2016 Carbon-based metal-free catalysts *Nat. Rev. Mater.* **1** 16064

[153] Lázaro M J *et al* 2015 Carbon-based catalysts: synthesis and applications *Comptes Rendus Chim* **18** 1229–41

[154] Hu C, Lin Y, Connell J W, Cheng H, Gogotsi Y, Titirici M and Dai L 2019 Carbon-based metal-free catalysts for energy storage and environmental remediation *Adv. Mater.* **31** 1806128

[155] Gaur V K, Gautam K, Sharma P, Gupta S, Pandey A, You S and Varjani S 2022 Carbon-based catalyst for environmental bioremediation and sustainability: updates and perspectives on techno-economics and life cycle assessment *Environ. Res.* **209** 112793

[156] Kundu D *et al* 2024 Heterogeneous catalysts for sustainable biofuel production: a paradigm shift towards renewable energy *Biocatal. Agric. Biotechnol.* **62** 103432

[157] Wang S, Yan W and Zhao F 2020 Recovery of solid waste as functional heterogeneous catalysts for organic pollutant removal and biodiesel production *Chem. Eng. J.* **401** 126104

[158] Havilah P R, Sharma A K, Govindasamy G, Matsaka L and Patel A 2022 Biomass gasification in downdraft gasifiers: a technical review on production, up-gradation and application of synthesis gas *Energies* **15** 3938

[159] Barker-Rothschild D, Chen J, Wan Z, Renneckar S, Burgert I, Ding Y, Lu Y and Rojas O J 2024 Lignin-based porous carbon adsorbents for CO_2 capture *Chem. Soc. Rev.* **54** 623–52

www.ingramcontent.com/pod-product-compliance
Lightning Source LLC
Chambersburg PA
CBHW080543220326

41599CB00032B/6343